12th International Workshop on Low Temperature Electronics 2016

Journal of Physics: Conference Series Volume 834

Tempe, Arizona, USA
18 – 21 September 2016

ISBN: 978-1-5108-4042-3
ISSN: 1742-6588

Printed from e-media with permission by:

Curran Associates, Inc.
57 Morehouse Lane
Red Hook, NY 12571

Some format issues inherent in the e-media version may also appear in this print version.

Printed by Curran Associates, Inc. (2017)

For permission requests, please contact the Institute of Physics at the address below.

Institute of Physics
Dirac House, Temple Back
Bristol BS1 6BE UK

Phone: 44 1 17 929 7481
Fax: 44 1 17 920 0979

techtracking@iop.org

Additional copies of this publication are available from:

Curran Associates, Inc.
57 Morehouse Lane
Red Hook, NY 12571 USA
Phone: 845-758-0400
Fax: 845-758-2633
Email: curran@proceedings.com
Web: www.proceedings.com

Table of contents

Volume 834

12th International Workshop on Low Temperature Electronics

18–21 September 2016, Tempe, AZ, USA

Accepted papers received: 7 April 2017
Published online: 2 May 2017

Preface

Papers

IOP Conf. Series: Journal of Physics: Conf. Series **834** (2017) 011001 doi:10.1088/1742-6596/834/1/011001

The present volume of the *Journal of Physics: Conference Series* represents contributions from participants of the 12th International Workshop on Low Temperature Electronics held in Tempe, Arizona, USA from September 18-21, 2016. The conference was organized by the School of Earth & Space Exploration at Arizona State University.

The International Workshop on Low Temperature Electronics (WOLTE) is a biennial conference devoted to the presentation and exchange of the most recent advances in the field of low temperature electronics and its applications. This international forum is open to everyone in the field.

The technical program included oral presentations and posters on fundamental properties of cryogenic materials, cryogenic transistors, quantum devices & systems, astronomy & physics instrumentation, and fabrication of cryogenic devices. More than 50 scientists and engineers from various academic, government, and industrial institutions in Europe, Asia, and the Americas attended the conference.

We would like to thank all speakers for their presentations and all attendees for their participation. We would also like to express our sincerest gratitude to our sponsors: Lake Shore Cryotronics, ASU NewSpace, ASU School of Earth & Space Exploration, and IRA A. Fulton Schools of Engineering for making this conference possible.

IOP Conf. Series: Journal of Physics: Conf. Series **834** (2017) 011002 doi:10.1088/1742-6596/834/1/011002

Peer review statement

All papers published in this volume of *Journal of Physics: Conference Series* have been peer reviewed through processes administered by the proceedings Editors. Reviews were conducted by expert referees to the professional and scientific standards expected of a proceedings journal published by IOP Publishing.

IOP Conf. Series: Journal of Physics: Conf. Series **834** (2017) 012001 doi:10.1088/1742-6596/834/1/012001

Low Temperature Characterization of Hole Mobility in Sub-14nm Gate Length $Si_{0.7}Ge_{0.3}$ Tri-Gate pMOSFETs

R. Laviéville[1], C. Le Royer[2], S. Barraud[2], G. Ghibaudo[1]

1 IMEP-LAHC, Univ. Grenoble Alpes, Minatec Campus, 38016 Grenoble
2 LETI-CEA, Univ. Grenoble Alpes, Minatec Campus, 38054 Grenoble

romain.lavieville@imep.grenoble-inp.fr

Abstract. We performed low temperature characterization of hole mobility in SiGe Trigate nanowires (NW) with gate length scaled down to 10nm. Trigate NWs with high-k/metal gate stack were fabricated on SOI wafers using solely optical lithography to design the wires with width down to 15nm. Drain current measurements are conducted within a cryogenic probe station enabling study on a 80K-350K temperature range. From these measurements we extracted the temperature dependence of the low field mobility μ_0 for a wide range of gate lengths and NW widths using the Y-function method in order to cancel the influence of the mobility attenuation factor due to series resistances. The impact of the temperature over the mobility can be used to identify the dominant scattering mechanism in the channel. Overall, we observed that hole transport is predominantly limited by the extra scattering due to neutral defects in all devices with a gate length under 40 nm. Source/Drain implantations trace the origins of these defects. Consequently, these particular process steps should gain special care in the design of further node generation as they critically hinder charge mobility at that scale.

1. Introduction

The successive inventions of Metal Oxide Semiconductor Field Effect Transistors (MOSFETs), Complementary Metal Oxide Semiconductor (CMOS) and the integrated circuits (ICs) in the 1960s initiated the pace of transistors scaling. In 1967, Gordon Moore observed that the number of transistors in a dense integrated circuit approximatively doubled every two years and this pace should continue for at least another decade. Moore's law has been unexpectedly respected for more than forty years by the microelectronic industries since it has irremediably led to cost reduction per fabricated transistor. It is only since 2012 that the transition time has slightly increased with the introduction of the 22 nm transistor node. Actually, since dimensions shrink, advanced fabrication and new physical phenomena challenge this development. Technological evolutions had to be implemented to reach adequate transistor performances by increasing their frequency operation and reducing their power consumption. The introduction of other channel material and strain engineering [1] in order to boost carrier mobility initially degraded by short channel effects illustrate such evolutions. Another recent advancement is the introduction of high-k dielectric in the metal gate stack at the 45 nm node in order to counter increasing gate leakage [2]. Recently, Tri-Gate MOSFET (as FinFET) has been introduced at the 22 nm transistor node by Intel and they now sell microprocessors with built-in 14nm transistor node. Thanks to their multigate geometry, which confers excellent electrostatic control over the channel, they provide a steeper sub-threshold slope, allowing operation at lower voltage to reduce

IOP Conf. Series: Journal of Physics: Conf. Series **834** (2017) 012001 doi:10.1088/1742-6596/834/1/012001

power and/or improve switching speed, resulting in better performance than their planar counterpart [3].

Channel mobility has always been used as a figure of merit in MOS technology since it characterizes well the carrier transport, representative of the transistor global performance. In this study we probed the hole mobility in strained SiGe channel in Tri-Gate pMOSFETs with alternative nanowire (NW) geometry. Such transistors are considered as promising candidates alongside FinFETs to continue scaling without performance loss. From these measurements we extracted the temperature dependence of the low field mobility μ_0 for a wide range of gate lengths and NW widths using the Y-function method. The impact of the temperature over the mobility can be used to identify the dominant scattering mechanism in the channel. Overall, we observed that hole transport is predominantly limited by the extra scattering due to neutral defects in all devices with a gate length under 40 nm.

2. Experiments and electrical parameters extraction methodology

2.1. SiGe Tri-Gate pMOSFET fabrication

Tri-gate pMOSFETs were produced at CEA-LETI on 300 mm strained Silicon Germanium on Insulator (SGOI) wafers. Our NW devices are fabricated along the <110> direction. The process flow in figure 1 details the major steps of our fabrication. We first produce the NW using a mesa isolation technique which consists of patterning transistor's channel width by deep UV photolithography, then trim the resist and etch the substrate before proceeding to successive steps of fin oxidation in order to reduce the diameter as small as ≈15 nm. Next, we deposit the high-k/metal gate stack before same photolithography process described above to pattern the gate and achieve gate length scaled down to 10nm. Moreover, lithography masks had broad range of gate and channel size in order to obtain different panel of devices. Thus, we obtain triple sided gate shape on all NWs. Then offset spacers are added on the sidewall of the gate before silicon raised source/drain, light dose implantation and annealing activation in order to reduce access resistances. Second spacer is added for more protection of the channel from a second highly dose source and drain implantation, activation spike anneal and silicidation. Finally, tungsten contacts and standard Cu back-end-of-the-line processes end the fabrication.

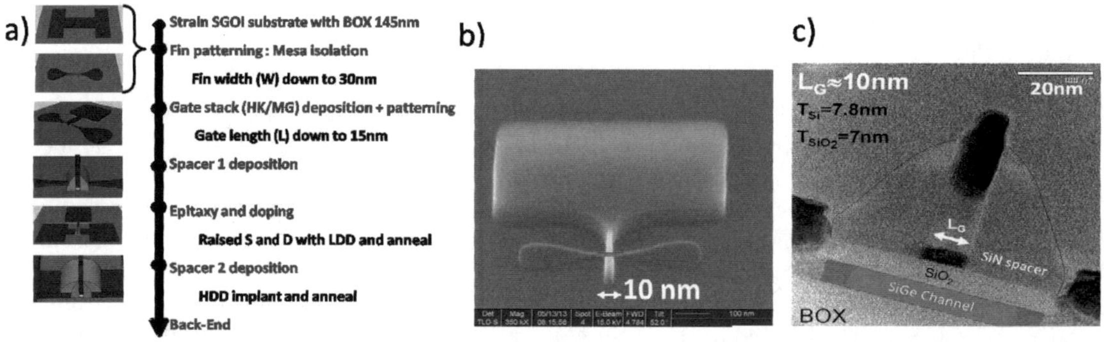

Figure 1 a) Process flow of Tri-gate Nanowire MOSFET. b) Top view SEM of a 10nm tri-gate length NW transistor after gate etching and c) TEM of a longitudinal cross-section of the same NW. On b) and c) is highlighted in red the effective channel.

Furthermore, along the single wire MOSFET depicted in figure 1b and 1c, multi wires MOSFETs have also been produced. They consist of ten parallel wires sharing a common Source, Drain and Gate (see figure 2). The multiplication of channels on a single device increases the drain current intensity allowing more accurate parameter extraction for mobility analysis.

IOP Conf. Series: Journal of Physics: Conf. Series **834** (2017) 012001 doi:10.1088/1742-6596/834/1/012001

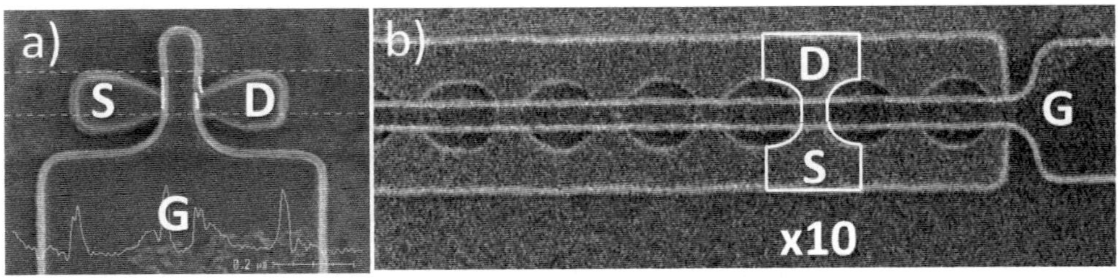

Figure 2 SEM of a) single wire and b) multi-wires tri-gate MOSFETs after gate patterning.

2.2. *Electrical characterization setup*
The electrical characterization was conducted on a manual cryogenic probe station under vacuum using triaxial cables and a semiconductor device analyzer (Agilent B1500). Results presented in this study range from 80K to 350K in order to preserve the devices from special quantum effects appearing usually below 50K [4]. Simple three probes measurements to contact the source, drain and gate of transistors were conducted to record Id(Vg) curves. We applied a small Voltage bias on the drain (Vd=20 mV) in order to stay in the linear region of these transistors as required by the Y-function methodology.

2.3. *Y-function methodology*
As for all MOS technology, Tri-gate MOSFETs have different mobility than bulk for same material. Its effective mobility is described by the electric field penetration into the channel, quantization in the inversion layer and scattering mechanisms and is modeled as follows:

$$\mu_{eff}(V_{GS}) = \frac{\mu_0}{1+\theta_1 V_{GT}+\theta_2 V_{GT}^{\ 2}} \tag{1}$$

where μ_0 is the low field mobility and $\theta_1$and θ_2 are respectively a linear and a quadratic attenuation factors due to scattering mechanisms.

Then we can express the linear current in a simplified equation at low transverse field as:

$$I_{Dlin}(V_{DS}, V_{GS}) = \frac{W.C_{ox}.\mu_0.V_{DS}.V_{GT}}{L_{eff}(1+\theta_1 V_{GT})} \tag{2}$$

where W is the width of the device, C_{ox} is the oxide capacitance and L_{eff} is the effective length of the channel. Note that the quadratic term attenuation θ_2 related to scattering occurring at the gate oxide/channel interface [5] is neglected as the gate voltage is restricted to 1.5 V. Then, Y-function is defined as the combination of the drain current over its squared root of transconductance as [6],

$$Y = \frac{I_D}{\sqrt{g_m}} = \sqrt{\frac{W.C_{ox}.\mu_0.V_{DS}}{L_{eff}}}.V_{GT} \tag{3}$$

By canceling the influence of the first order mobility attenuation factors, $\theta_1 = \theta_0 + (R_{sd}.WC_{ox}\mu_0/L_{eff})$, R_{sd} being the source/drain series resistance, we are able to extract the low-field mobility μ_0 from the slope of the Y function provided W, C_{ox} and L_{eff} are known, while the threshold voltage V_{th} can be read at the x axis intercept. Typical plots of drain current and corresponding Y-function with gate voltage are shown in figure 3.

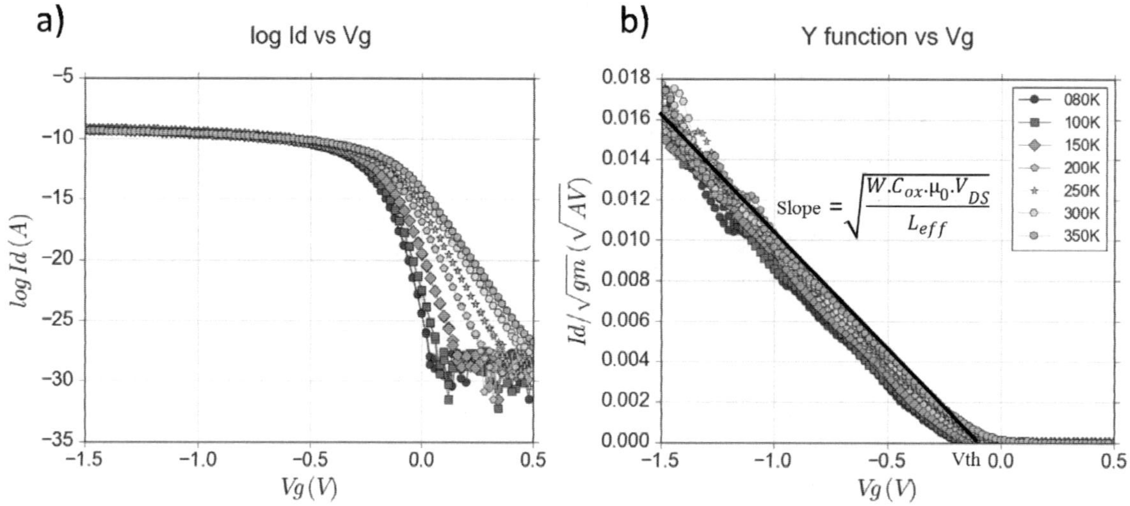

Figure 3 a) Typical semi-log Id-Vg curves showing subthreshold slope temperature dependence of Tri-gate pMOSFET (W=220nm, L=25nm). b) Example of μ_0 extraction at 350K.

3. Results and discussions

3.1. Results

We applied the Y-function methodology in order to extract μ_0 from various devices for a temperature range from 80K to 300K. We report here representative devices with several gate lengths from 400 nm down to 10 nm for fixed 15 nm, 60 nm and 240 nm channel widths. Graphics in figure 4a) display low-field mobility behavior versus gate length at each width. Moreover, each curve within a graphic represents the characteristic of μ_0 at a given temperature. From these data we observe two global trends. Firstly, the low-field mobility decreases along the gate length at the same amplitude for all three devices, meaning only physical phenomena due to variation of gate length are responsible for this mobility attenuation. Secondly, lower temperatures reduce slightly the scattering of the carriers enhancing the mobility in the channel.

In order to interpret the reduction of mobility with smaller gate length, we plotted the same data with different axes where the low-field mobility is now represented in function of temperature (see figure 4b)). Again, we observe similar trend for the low field mobility for each nanowire width (excepted when W=60 nm and L=100 nm) which depend now on the gate length. For long gate lengths over 40 nm, the mobility increases when the temperature decreases, which is a typical trend for acoustic phonon scattering. However, for gate lengths below 40 nm, the mobility is nearly independent of temperature, which is characteristic of neutral impurity scattering. Thanks to Matthiessen's rule it is possible to express the mobility within the channel as the sum of its various components in function of temperature as [7, 8]:

$$\frac{1}{\mu_0(T)} = \frac{T}{300.\mu_{ph}} + \frac{300}{\mu_c.T} + \frac{1}{\mu_{neu}} \quad (4)$$

where μ_{ph} is the mobility contribution related to the acoustic phonon, μ_c is related to remote coulombic interactions while μ_{neu} is an additional term due to neutral defects [9,10]. Temperature dependences shown in figure 4b) are well fitted with equation (4).

Thus we were able to extract the different components of the mobility for each device. Figure 5 displays the percentage of contribution to low-field mobility scattering from each of its extracted component in respect to the gate length. From these graphics it is clear that scattering due to neutral defects in the channel prevails over acoustic phonon and coulombic interactions at ultimate gate scaling. For 15nm nanowire width and gate length of 10 nm, neutral defects scattering represents over 90% of mobility attenuation. Its contribution fades for longer gate length but holds majority for narrower wires, when for wider devices (W=240 nm) we get back to acoustic phonon-dominated scattering.

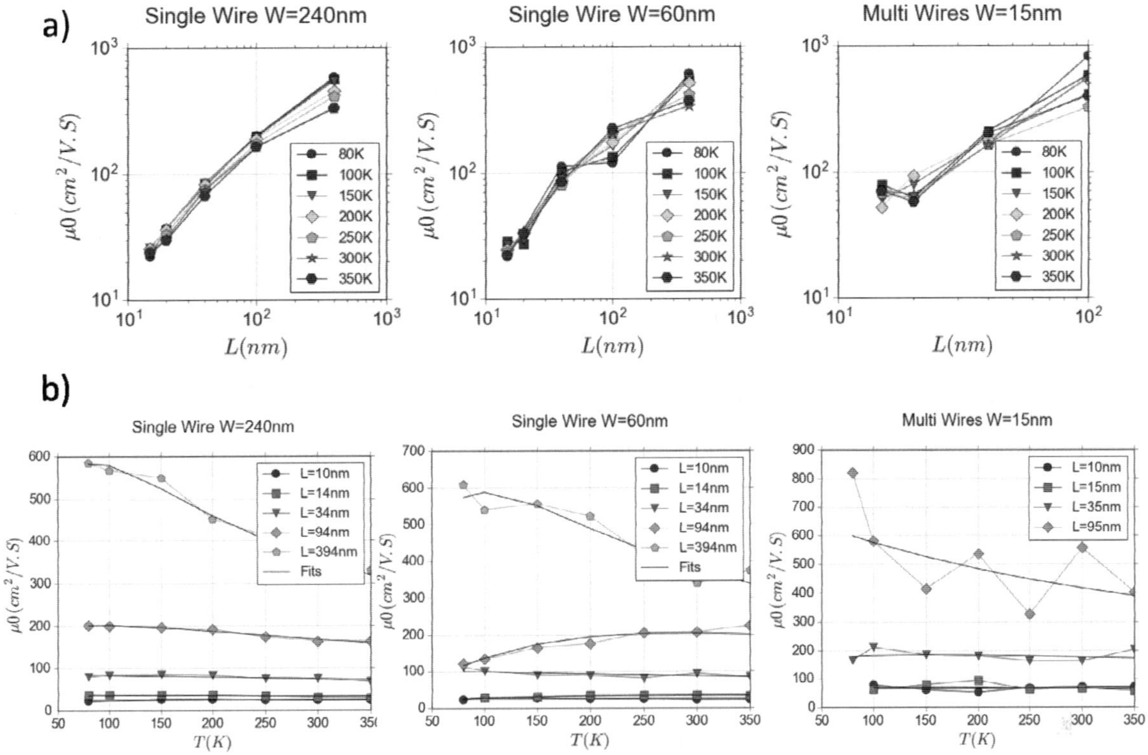

Figure 4 a) Low-field mobility μ_0 extracted with Y-function in function of gate length. Within a graph each curve represents μ_0 behavior for a distinctive temperature. **b)** μ_0 in function of temperature for different gate lengths. All fits let us to extract the different mobility components at each gate length.

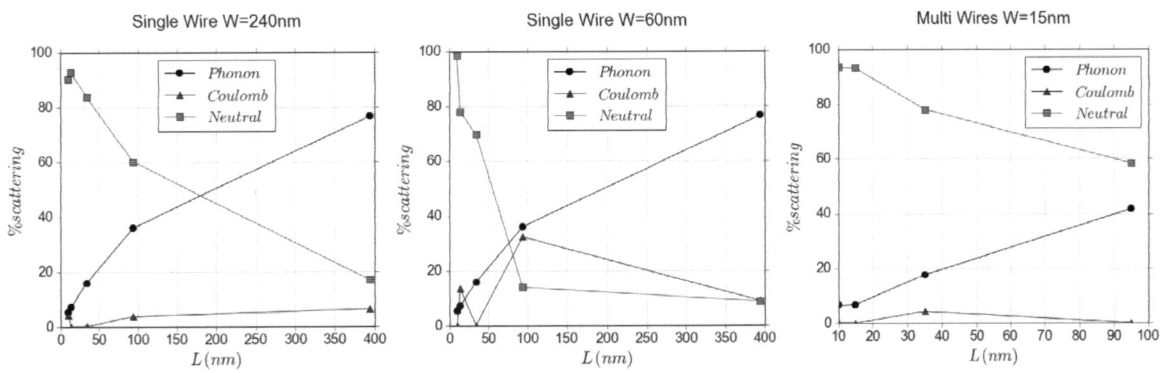

Figure 5 Percentage of contribution to scattering from each mobility component in respect to the gate length.

3.2. Discussions

As shown in the previous paragraph, neutral defects in short channel wires dominates scattering considerably over acoustic phonon and coulombic interactions. Origin of these defects are subjected to interpretations, thus we expose here our understanding of the phenomenon. We believe these defects are introduced after implantation of doping of the source and drain areas. At this stage of the fabrication, the channel under the gate is protected by a first set of spacers at both sides of the gate. Thus, the ions and interstitial atoms created in the SiGe crystal by the implantation are believed to be injected from the source and drain into the channel ends. Naturally, these devices are still in development and will need further adjustments to match industrial need. Improvement of low-field mobility for smaller devices should arise from better controlled doping and annealing steps during fabrication, in particular in-situ doped source/drain process without implantation could help in this respect. As an alternative, spacer size could also be adjusted to reduce the injection of these defects into the channel. Moreover, at this development stage, the temperature-independent mobility behaviour of our devices induces cooling will not enhance their electrical properties, resulting in low current intensity at any temperature.

4. Conclusions

In this work we characterized hole mobility within the channel of $Si_{0.7}Ge_{0.3}$ Tri-gate pMOSFETs over a 80K - 350K temperature range. Hole mobility from various devices in gate lengths and widths is extracted using the Y-function method in order to remove the attenuation from the source and drain series resistances. Then, the influence of temperature over the low-field mobility is used to identify the dominant scattering mechanisms in the channel. In particular, we discriminated the scattering components using Matthiessen's rule into phonon, coulombic and neutral defects scattering mechanisms. We identified strong mobility attenuation due to neutral defects in all devices from 40 nm gate length and below, regardless of their width, due to independent temperature behavior. Specifically for 15 nm gate length devices, neutral defects scattering represents over 90% of mobility attenuation. We believe Source/Drain implantations trace the origins of these defects. Therefore, these process steps should gain special care in the design of further node generation as they limit charge mobility at that scale.

Acknowledgements

This work was supported by Compose 3, a project from the European framework. The authors also thank LETI clean room facilities for providing samples and Xavier Mescot for technical assistance in cryogenic experiment.

References

[1] Takagi S, Hoyt J-L, Welser J-J and Gibbons J-F Comparative study of phonon-limited mobility of two-dimensional electrons in strained and unstrained Si metal-oxide-semiconductor-field-effect-transistors 1996 *J. Appl. Phys* **80** 1567

[2] Wallace R-M and Wilk G-D Materials issues in high-k gate dielectric selection and integration 2005 *High Dielectric Constant Materials: VLSI MOSFET applications* (Springer Berlin Heidelberg).

[3] Auth C, Allen C, Blattner A, Bergstrom D, Brazier M,Bost M, Buelher M, Chikarmane V, Ghani T, Glassman T, Grover R, Han W, Hanken D, Hattendorf M, Hentges P, Heussner R, Hicks J, Ingerly D, Jain P, Jaloviar S, James R, Jones D, Jopling J, Joshi S, Kenyon C, Liu H, McFadden R, McIntyre B, Neirynck J, Parker C, Pipes L, Post I, Pradhan S, Prince M, Ramey S, Reynolds T, Roesler J, Sandford J, Seiple J, Smith P, Thomas C, Towner D, Troeger T, Weber C, Yashar P, Zawadzki K and Mistry K A 22nm high performance and low-power CMOS technology featuring fully-depleted tri-gate transistors, self-aligned contacts and high density MIM capacitors 2012 *Symposium on VLSI Technology (VLSIT)*

IOP Conf. Series: Journal of Physics: Conf. Series **834** (2017) 012001 doi:10.1088/1742-6596/834/1/012001

[4] Lavieville R, Triozon F, Barraud S, Corna A, Jehl X, Sanquer M, Li J, Abisset A, Duchemin I and Niquet Y-M Quantum Dot Made in Metal Oxide Silicon-Nanowire Field Effect Transistor Working at Room Temperature 2015 *Nano Lett*, **15**, 2958.

[5] Ong T-C, KO P-K and Hu C, gate-Oxide MOSFET's at 77 K 1987 *IEEE Trans. Electron Devices* **34**, 2129

[6] Ghibaudo G, New method for the extraction of MOSFET parameters. 1988 *Electron Lett* **24**(9):543-5

[7] Cros A, Romanjek K, Fleury D, Harrison S, Cerutti1 R, Coronel1 P, Dumont B, Pouydebasque A, Wacquez R, Duriez B, Gwoziecki R, Boeuf F, Brut H, Ghibaudo G and Skotnicki T, Unexpected mobility degradation for very short devices : A new challenge for CMOS scaling. 2006 *IEEE Proc. IEDM*, 439-402

[8] Shin M, Shi M, Mouis M, Cros A, Josse E, Kim G-T, Ghibaudo G, Low temperature characterization of mobility in 14 nm FD-SOI CMOS devices under interface coupling conditions. 2015 **108** 30-35

[9] N. Sclar. Neutral Impurity Scattering in Semiconductors . 1956 *Phys. Rev.,* **104**, 6, 1559–1561

[10] T.C. McGill et R. Baron. Neutral impurity scattering in semiconductors. 1975 *Physical Review B*, **11**, 5208–5210

IOP Conf. Series: Journal of Physics: Conf. Series **834** (2017) 012002 doi:10.1088/1742-6596/834/1/012002

BSIM3 parameters extraction of a 0.35 µm CMOS technology from 300K down to 77K

Laurent Varizat[1,2], Gerard Sou[1] and Malik Mansour[2]

[1] Electrical and Electromagnetism Laboratory, Pierre and Marie Curie University, 4 Place Jussieu, 75252 Paris Cedex 05, France

[2] Laboratory of Plasma Physics, CNRS, Ecole Polytechnique, Route de Saclay, 61128 Palaiseau, France

E-mail: laurent.varizat@upmc.fr

Abstract. In this work, we present a commercial 0.35 µm/3.3 V CMOS technology behaviour study of both linear and gate-enclosed transistors from room temperature down to 77 Kelvin. Cryogenic setup used to complete this study is first described. Measurement results are then discused and a model based on a BSIM3 parameters extraction is proposed.

Keywords: Cryogenic temperature MOSFET; BSIM3 model; Gate-enclosed layout.

1. Introduction

At low temperature, MOSFETs behaviour is modified. Gain and charge carrier mobility modifications as well as thermal noise and parasitic effect reductions are observed. Accurate knowledge of transistors behaviour is essential to design fully integrated instrumentation electronics and to achieve the best performances of integrated circuit in many fields like space instrumentation or nuclear science. Unfortunately, simulation models available from industrial foundries are meant to work only from $-40\,^{\circ}$C to $150\,^{\circ}$C. Previous works have investigated technology behaviours at low temperature. For instance, in [1] an evaluation of AMS 0.35 µm CMOS technology to be implemented in Mars landers is presented but the study is not going below 165K. However the knowledge of the transistors behaviour at low temperature is required for many other space applications for wich cryogenic environment can help improving performances [2, 3] by:

- Increasing speed due to the increase charge transport properties
- Reducing interconnect resistance
- Decreasing the thermal electrical noise
- Reducing parasitic effects and thus improving reliability regarding latch-up, leakage currents of electromigration
- Increasing gain

2. Setup and methods

A dedicated test bench has been developed to caracterize electrical transistors properties at cryogenic temperature. A test device composed of both PMOS and NMOS transistors with different sizes and geometries was designed. Table (1) lists the transistors geometries indexed into four main domains.

Type	Size (W x L)		Layout architecture
NMOS	50 µm	x 0.35 µm	Linear
	50 µm	x 10 µm	Linear
	1000 µm	x 10 µm	Linear
	50 µm	x 0.35 µm	Gate-enclosed
	50 µm	x 10 µm	Gate-enclosed
	1000 µm	x 10 µm	Gate-enclosed
PMOS	50 µm	x 0.35 µm	Linear
	50 µm	x 10 µm	Linear
	1000 µm	x 10 µm	Linear
	50 µm	x 0.35 µm	Gate-enclosed
	50 µm	x 10 µm	Gate-enclosed
	1000 µm	x 10 µm	Gate-enclosed

Table 1: List of transistors used for characterisation.

For each size, both linear and gate-enclosed transistor geometries have been studied. Gate-enclosed transistors [4] show a better resistance to ionising radiation which is an asset for space applications. This design removes gate's edges and reduces leakage currents due to the bird's beak effect [5]. Figure (1a) shows the layout of a linear NMOS 50 µmx10 µm while figure (1b) shows the equivalent gate-enclosed transistor.

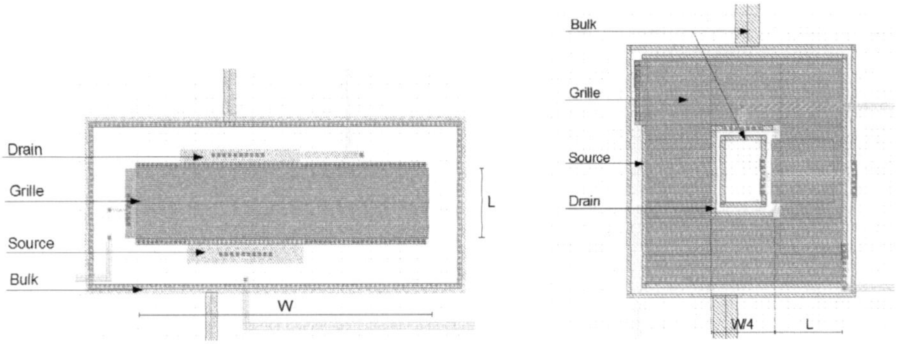

(a) (b)

Figure 1: Two types of 50 µm x 10 µm NMOS transistor's layout : (a) Linear layout, and (b) Gate-enclosed layout.

2.1. Cryogenic setup

Figure (2) shows the cryogenic test bench. The circuit under test is inserted inside a cryostat immersed in a liquid nitrogen tank. Vacuum is maintained inside the cryostat to prevent air condensation and freeze on the device. A metallic frame is used to keep the cylinder in place during measurements. The cryostat immersion depth in the liquid nitrogen tank is controlled thanks to a micro-controller and a PT100 thermistor positioned on the device side. During

IOP Conf. Series: Journal of Physics: Conf. Series **834** (2017) 012002 doi:10.1088/1742-6596/834/1/012002

measurement the desired temperature is set by moving the cryostat up and down. This setup allows to do measurements at different temperatures from room temperature down to 77K.

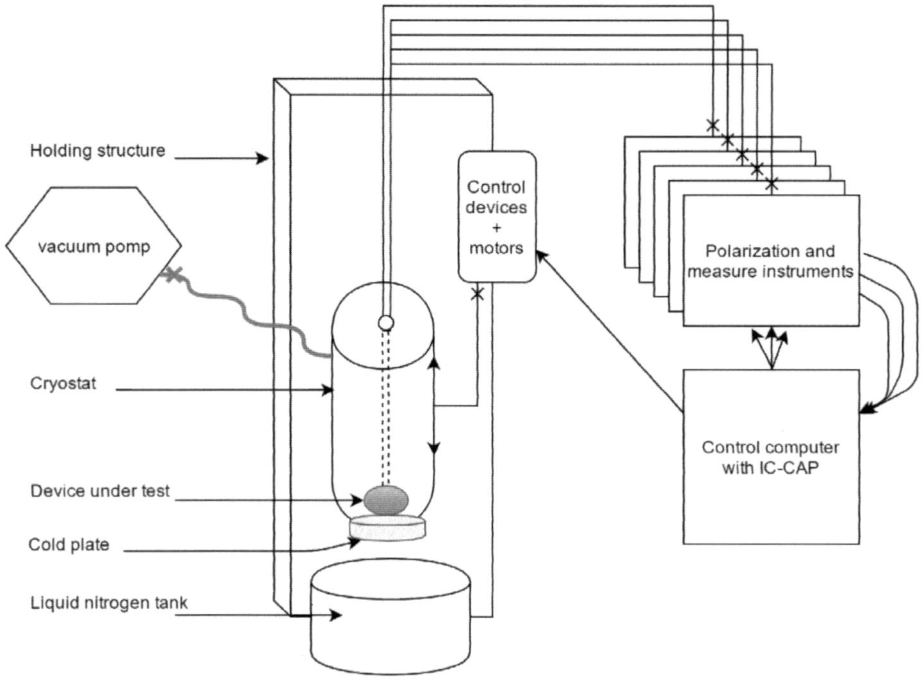

Figure 2: Cryogenic test bench.

2.2. Extraction tools

A DC generator is used to supply the device while a semiconductor analyser[1] performs current-voltage and capacitive measurements. Capacitive measurements shown on figure (3) allow data extraction of following parameters:

1. BD's junction;
2. Gate oxide;
3. Intrinsic capacities;
4. Overlap effect;

A voltage sweep is performed from -0.3V to 3.6V with a 0.1V step between *Hi* and *Lo* terminals (fig (3)). All instruments are controled by IC-CAP[2] software controls. Measurement points are corrected from the parasitic resistances of the drain and the source according to the temperature [6]. BSIM3v3 model parameters are extracted for each transistor category with IC-CAP using a precise extraction flow. An optimisation phase is then performed to improve the model's precision using Levenberg-Marquart algorithm which presents the best performance compared to other available algorithms.

3. Thermal study

To obtain a valid model describing the transistors behaviour around 100K, characteristics have been measured at 77K, 100K, 150K, 200K and 300K as a comparison.

[1] Keysight™ B1500A Semiconductor Device Analyser

[2] Keysight™ Technologies : `http://www.keysight.com/`

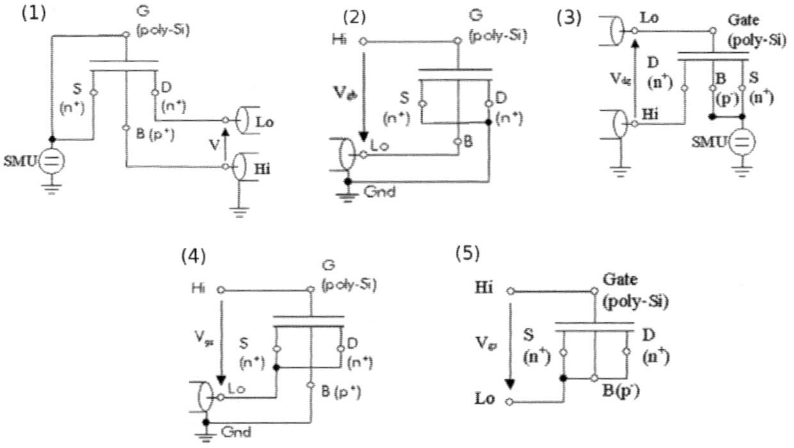

Figure 3: CV measurement representation schematic.

3.1. Current-voltage measurements

Foundry model is meant to work from −40 ° C to 150 ° C. Figure (4) shows a transistor drain current temperature dependency for a given size and polarisation. On the same graph are plotted the foundry model simulation, the measurements and the extracted model. Below −40 ° C a divergence between measurement and foundry model is observed. This divergence is growing as the temperature decrease so that around 100K the foundry model can not be used anymore.

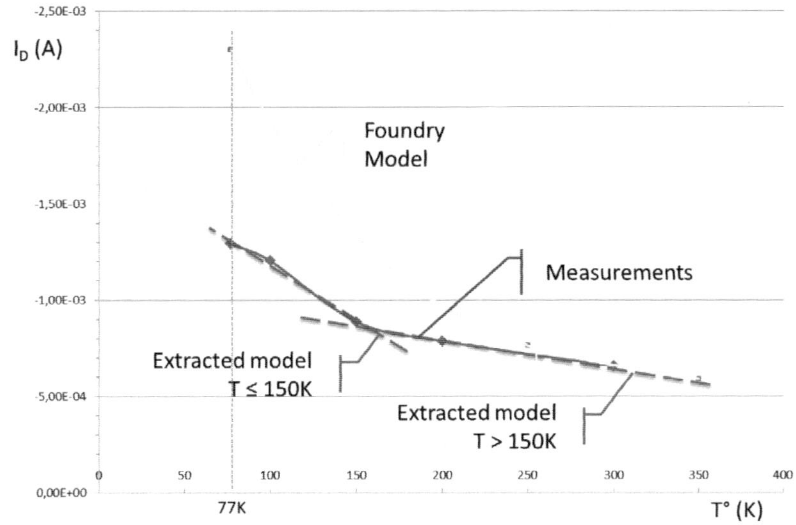

Figure 4: Extracted model simulation compared to foundry model and measurements for a 50 µm × 10 µm with $\{V_{GS};V_{BS};V_{DS}\}$={3.3V; 0V; 3.3V}.

One of the most important temperature dependant parameter is the threshold voltage.

Threshold voltage V_{th_0} defines the gate polarisation voltage split between two transistors states. Before threshold voltage, the transistor is in off mode, corresponding to a low current from drain to source. Conversely, after threshold voltage, the transistor is in linear or saturate mode. Table (2) shows the threshold voltage variation with temperature. The absolute value of the threshold increases as the temperature decreases. This implies that for cryogenic temperature, the lowest transistor polarisation has to be greater than at room temperature. For PMOS transistors, a drift of half a volt is observed from room temperature down to 77K and reaches the value of -1.2V.

	300K	200K	150K	100K	77K
NMOS	0.5V	0.58V	0.66V	0.72V	0.76V
NMOS HR		0.39V	0.4V	0.49V	
PMOS	-0.7V	-0.9V	-1.0V	-1.2V	-1.2V
PMOS HR		-0.86V	-0.89V	-1.1V	

Table 2: V_{th_0} temperature dependence for different types of transistors.

Furthermore, silicon conductance is modified and evolves conversely to the temperature as shown on figure (5) by the output conductance curves, g_{ds}, which quantify the drain current variation with a drain-source voltage variation while keeping the gate-source voltage constant (1).

$$g_{ds} = \left.\frac{\partial I_D}{\partial V_{DS}}\right|_{V_{GS}} \tag{1}$$

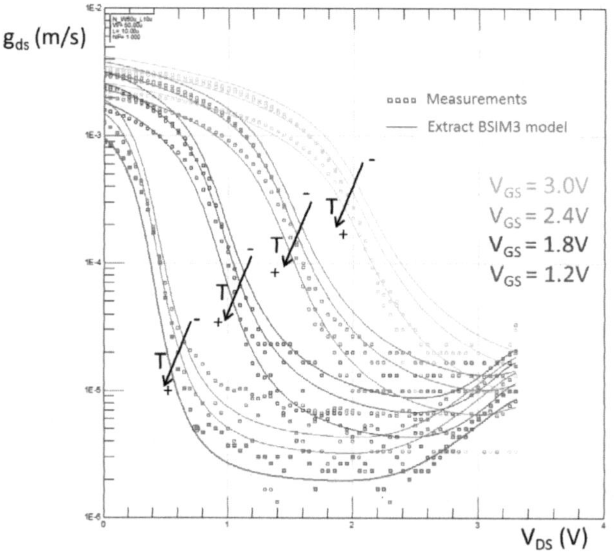

Figure 5: G_{DS} = f(V_{DS}, T) for a 50 µm × 10 µm NMOS.

3.2. Capacitive measurements

Capacitive measurements are useful to extract physical parameters of BSIM3's model. Figure (6) shows the capacity between the gate and drain-source-bulk in function of its polarisation. Besides a slight offset due to V_{th} drift no significant difference in capacitive measurements between 300K and 77K is observed.

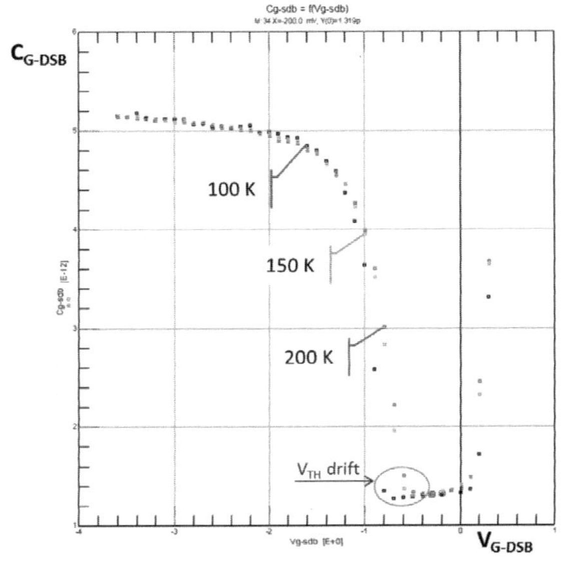

Figure 6: C_{G-DSB} temperature dependence for V_{G-DSB} sweep for a gate-enclosed 50 µm × 10 µm PMOS transistor.

3.3. *Physical interpretation: freeze-out effect*

The diminution of temperature provokes an increase in electron's mobility which is observed by an increase of the drain current. In the meantime, there is a diminution on charge carrier density with temperature [7]. When the density is too low, a drop in transistors current is observed. This well known freeze-out effect corresponds to a charge carrier gel and depends on the doping concentration. Figure (7) shows the electron density according to the temperature. The dots correspond to the electron density and the Fermi energy at 77K. The electron density decreases about two order of magntitude from 300K to 77K. As a result the current flowing in transistors decreases rapidly according to the temperature.

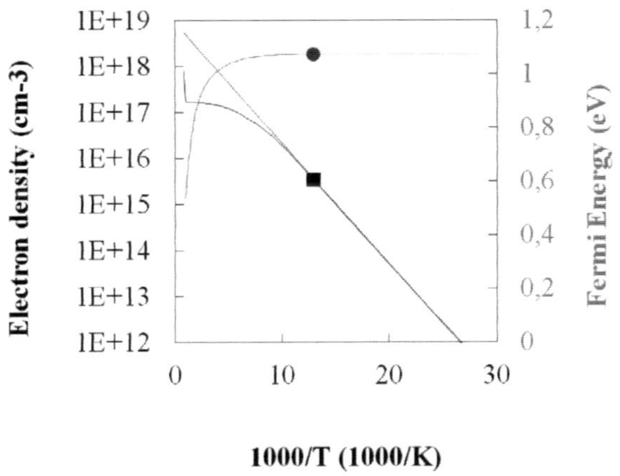

Figure 7: Electron density evolution with temperature and the corresponding Fermi energy.[3]

IOP Conf. Series: Journal of Physics: Conf. Series **834** (2017) 012002 doi:10.1088/1742-6596/834/1/012002

4. BSIM3 parameters extraction

BSIM3 model [8] does not take in account low temperature physical effect like the freeze-out effect. Whereas some works present new models in order to take the freeze-out effect in account [2,9] we have adjusted BSIM3 physical parameters to bypass this lack. Figure (8) shows $I_D V_G$ and $I_D V_D$ curves between measurements and simulations based on extracted model at 100K. Both simulations and measurements are in good agreement.

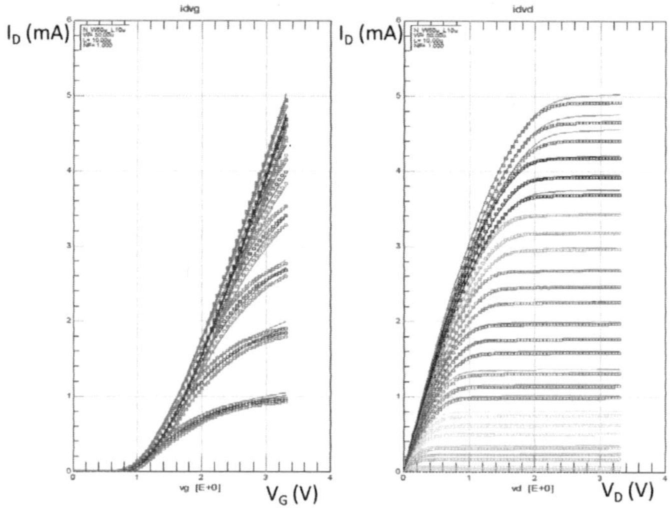

Figure 8: $I_D V_G$ and $I_D V_D$ measurements (dots) and extracted model simulations (lines) from a standard 50 µm × 10 µm NMOS at 100K.

5. Conclusion

We presented a cryogenic study of a commercial standard 0.35 µm bulk CMOS technology for both normal and gate-enclosed layout between 77K and 300K. For this purpose, an integrated circuit with transistors, resistors and capacitors of several dimensions and geometries has been designed, as well as a dedicated cryogenic set-up. Current-voltage capacitive measurements and parameters extraction have been performed using IC-CAP (Keysight^{TM}) software. As expected, a deviation, growing with decreasing temperature between measurements and foundry based models simulations is observed and is being assigned to physical phenomena, like freeze-out effect. Around 100K, deviation is such that models become unusable for a large variety of integrated circuit design. A model based on a BSIM3 parameters extraction has been proposed. Ready to use Cadence libraries have been developed. Librairies are suitable for the accurate design of cryogenic circuits in a wide temperature range from 300K down to 77K and for given transistor sizes.

References

[1] Ramos-Martos J, Ragel-Morales A, Ceballos-Cceres J, Mora-Gutirrez J M, Lagos-Florido M A and Espejo-Meana S 2012 URL `http://digital.csic.es/handle/10261/84095`
[2] Zhao H and Liu X 2014 *Cryogenics* **59** 49–59 ISSN 00112275 URL `http://linkinghub.elsevier.com/retrieve/pii/S0011227513000969`
[3] Yangbo Y, Zhe G and Haisong L 2009 *Chinese J. Electronics* **18** 215

[3] adapted from Bart Van Zeghbroeck equations (1997)

[4] Fan X, Li P, Li W, Zhang B, Xie X, Wang G, Hu B and Zhai Y 2011 *Journal of Semiconductors* **32** 084002 ISSN 1674-4926 URL http://stacks.iop.org/1674-4926/32/i=8/a=084002?key=crossref.0b69fd2af8f7ede9b5f2a54a0c1cb0bf

[5] Oldham T R and McLean F B 2003 Total Ionizing Dose Effects in MOS Oxides and Devices URL http://ntrs.nasa.gov/search.jsp?R=20030032300

[6] Desai P D, James H and Ho C Y 1984 *Journal of physical and chemical reference data* **13** 1131–1172

[7] Street R A 1982 *Physical Review Letters* **49** 1187–1190 URL http://link.aps.org/doi/10.1103/PhysRevLett.49.1187

[8] Sheu B J, Scharfetter D L, Ko P K and Jeng M C 1987 *IEEE Journal of Solid-State Circuits* **22** 558–566

[9] Jia K, Sun W and Shi L 2011 *Journal of Semiconductors* **32** 064002 ISSN 1674-4926 URL http://stacks.iop.org/1674-4926/32/i=6/a=064002?key=crossref.e3cfc1a8b488df63b355f9938e0294db

IOP Conf. Series: Journal of Physics: Conf. Series **834** (2017) 012003 doi:10.1088/1742-6596/834/1/012003

Fast, high-fidelity readout of multiple qubits

N T Bronn, B Abdo, K Inoue, S Lekuch, A D Córcoles, J B Hertzberg, M Takita, L S Bishop, J M Gambetta, J M Chow
IBM T.J. Watson Research Center, 1101 Kitchawan Rd, Yorktown Heights, NY, 10598, USA
E-mail: `ntbronn@us.ibm.com`

Abstract. Quantum computing requires a delicate balance between coupling quantum systems to external instruments for control and readout, while providing enough isolation from sources of decoherence. Circuit quantum electrodynamics has been a successful method for protecting superconducting qubits, while maintaining the ability to perform readout [1, 2]. Here, we discuss improvements to this method that allow for fast, high-fidelity readout. Specifically, the integration of a Purcell filter, which allows us to increase the resonator bandwidth for fast readout, the incorporation of a Josephson parametric converter, which enables us to perform high-fidelity readout by amplifying the readout signal while adding the minimum amount of noise required by quantum mechanics, and custom control electronics, which provide us with the capability of fast decision and control.

1. Introduction

Quantum computers have the potential to solve certain problems much more efficiently than classical computers [3, 4, 5]. In 1994, Peter Shor proposed a quantum algorithm to factor large composite numbers into their respective primes that is exponentially more efficient than any known classical algorithm [6]. This ignited additional interest into quantum computing, with more experimental effort into constructing the basic building blocks. The basic unit of information for a quantum computer is a quantum bit (qubit), which physically can be any two-level quantum mechanical system. The challenge is to control and measure qubits, while prohibiting unwanted interactions with their environment, and scaling them up to larger systems. Currently, the largest qubit systems are with trapped ions or superconducting circuits [7, 8, 9, 10].

2. Qubit Control

At IBM we use the *transmon* style of superconducting qubit, consisting of a single $\mathrm{Al/AlO}_x\mathrm{/Al}$ Josephson junction shunted by a large capacitor [11]. The Josephson junction provides a nonlinear inductance which, when combined with the capacitive shunt, forms an an anharmonic oscillator. In such a system, the lowest two energy levels can be used as the qubit and can be uniquely addressed as the $|0\rangle$ and $|1\rangle$ computational states. Qubit control is provided by applying microwave frequency pulses at this transition frequency, typically $\sim 4-6$ GHz, which rotates the qubit state around the Bloch sphere about an axis that lies in the xy-plane (Fig. 1a). Applying a pulse that rotates the state by 180° would then take the $|0\rangle$ state to the $|1\rangle$ and vice versa. Thus, it is equivalent to the classical `NOT` gate in digital logic. If instead a 90° pulse is applied, the qubit state would be in an equal superposition of both $|0\rangle$ and $|1\rangle$. The phase of this superposition is set by the phase of the applied microwave-frequency tones. Because the

IOP Conf. Series: Journal of Physics: Conf. Series **834** (2017) 012003 doi:10.1088/1742-6596/834/1/012003

measurement of qubits projects the state into the z-axis, measuring this state would yield $|0\rangle$ in half of the cases and $|1\rangle$ in the other half.

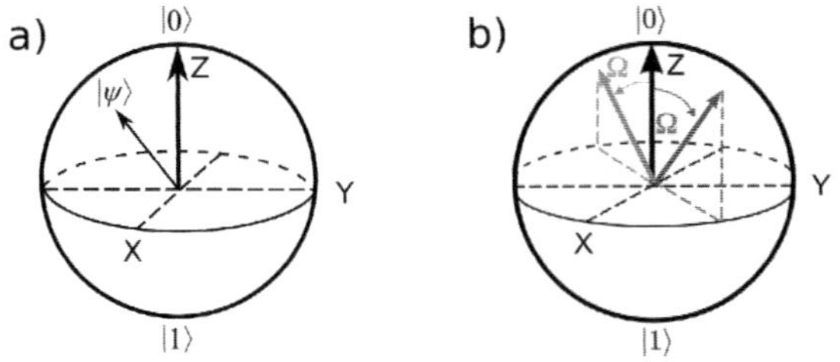

Figure 1. a) The qubit states $|0\rangle$ and $|1\rangle$ are represented on the Bloch sphere as the North and South poles, respectively. Any other point on the surface of the sphere represents a superposition of $|0\rangle$ and $|1\rangle$, with the equator being an equal superposition. b) Bloch sphere of the target qubit showing the dependence of its rotation direction on the state of the control qubit when the cross resonance operation is applied.

In order to complete a universal computational set of operations for quantum computing, there must also be operations between pairs of qubits. Such two-qubit gates can generate quantum entanglement between qubits, and the particular physical implementation we use is known as *cross-resonance gates* [12, 13, 14]. In a cross-resonance operation, a target qubit is driven via a control signal applied to a separate coupled qubit, known as the control qubit. When the two qubits are weakly coupled, such a microwave cross-resonance gate can result in a rotation of the target qubit that depends on the state of the control qubit (see Fig. 1b). Through proper adjustment of the target and control qubit microwave drive fields, it is possible to tune up a controlled-not (`CNOT`) entangling gate.

3. Qubit Readout

One particular architecture for realizing superconducting quantum processors is circuit quantum electrodynamics (QED), in which superconducting qubits are strongly coupled to microwave resonators [1, 2]. Circuit QED is a fantastic testbed for studying quantum optics and strong light-matter interactions, yet it also serves as a technique by which qubits can be controlled and read out effectively, while helping to isolate them from the outside environment. Superconducting qubits can be coupled to coplanar waveguide (CPW) resonators for readout and control, and also to one another, as shown in Fig. 2a. In the dispersive regime of qubit-resonator coupling the resonator frequency becomes qubit-state dependent, and thus through probing this resonator it becomes possible to discern the qubit state. Often we apply a readout tone at a frequency between the resonator frequencies corresponding to the $|0\rangle$ and $|1\rangle$ qubit states, and measure a state-dependent phase shift of the reflected readout tone. This type of measurement also has the advantage of leaving the qubit in the state it was measured, known as a *quantum non-demolition* (QND) measurement.

Figure 3 shows a standard readout chain used for the readout of superconducting qubits in a dilution refrigerator setup. Please note that the drawing does not show the attenuation and filtering configuration that is present on the input line for simplicity. In this scheme, the readout pulse propagates from left to right. It is first routed to a qubit/resonator system via a commercial three-port microwave circulator. The readout pulse interacts with the qubit/resonator system, reflects off the readout resonator, and carries information about the qubit state in its phase and amplitude. This readout signal is routed again, by the same three-port circulator, towards the output chain, where it passes though two wide-band isolators and a low-pass filter before it is amplified, first at the 3 K stage, by a high electron mobility transistor (HEMT) amplifier and

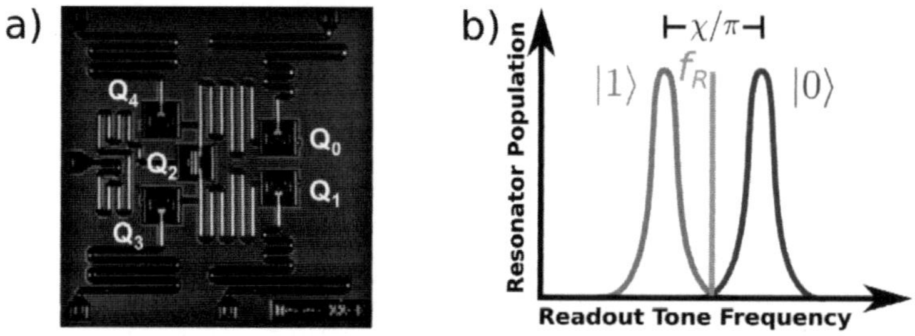

Figure 2. a) Micrograph of a five qubit processor, consisting of two bus resonators connecting the qubits and five readout resonators which are also used for control. b) Diagram of the qubit state dependence of the resonator, separated by the dispersive shift χ/π. The readout frequency f_R is typically set between the resonator's frequencies corresponding to the $|0\rangle$ and $|1\rangle$ states.

later at room temperature by commercial radio frequency (RF) amplifiers. Then, the amplified signal gets mixed down with a local oscillator (LO) to a low intermediate frequency (IF) on the order of tens of megahertz, digitized, and processed by a computer. We will repeatedly refer back to this readout chain in our discussion about improving it.

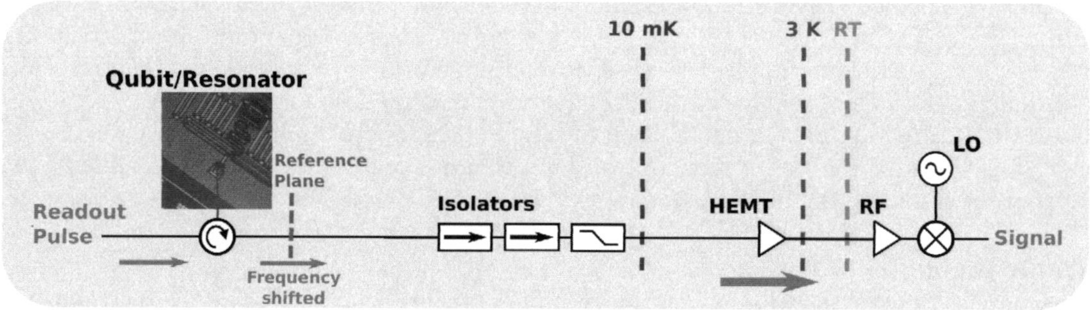

Figure 3. Standard readout scheme. A readout pulse is applied to the superconducting qubit/resonator system through an input line in the dilution refrigerator. The input line (not shown) includes several fixed attenuators and filters that are well-thermalized at different stages of the dilution refrigerator. The role of these attenuators and filters is to heavily attenuate the readout signal and noise coming from the room temperature instruments down to the level of a few microwave photons. This attenuated readout pulse is incident upon a microwave circulator that directs it to the qubit/resonator system, where it is reflected with a phase shift depending on the state-dependent resonance frequency of the resonator (we will refer signals back to the reference plane depicted here). The pulse then passes through two isolators and a low-pass filter, used to protect the device from noise coming down the output chain. The pulse is then amplified by a HEMT amplifier at the 3 K stage and further by room temperature RF amplifiers before it is mixed down with an LO to an IF and digitized.

4. Towards Fast, High-fidelity Readout

Fast, high-fidelity readout is crucial for building a quantum computer. While the standard scheme discussed thus far is sufficient to perform experiments in which measurements may be averaged over many rounds, carrying out more sophisticated experiments on a quantum computer which employ feedback and dynamic state-dependent operations is more challenging and require fast, high-fidelity readout [15, 16, 17]. For superconducting qubits, such a capability can be realized by introducing three components: a Purcell filter that allows increased resonator bandwidth without adversely affecting the qubit lifetime, a Josephson parametric converter, which operates at the quantum limit (i.e., adds the minimum amount of noise required by quantum mechanics), for amplifying the readout tone, and logic and control electronics that facilitate quick measurement determination and feedback. The following sections describe each of these in further detail.

4.1. Purcell Filter

In standard circuit QED, one limitation on qubit lifetime, T_1, is relaxation via spontaneous emission of a photon by the resonator, known as the Purcell effect. In this limitation, the qubit lifetime is inversely proportional to the qubit-resonator coupling strength and readout resonator bandwidth, κ. In particular, approximating the qubit as a harmonic oscillator, its spontaneous emission-limited lifetime is given by a generalized time constant $T_1^{\text{Purcell}} = C_q/\text{Re}[Y_q]$, where C_q is the capacitance of the qubit and Y_q is the admittance seen by the qubit at its transition frequency. In the case of standard QED, Y_q corresponds to the frequency-dependent admittance seen by the qubit, which depends on the readout resonator frequency and κ. To realize fast readout, κ should be large, so as to allow photons that carry information about the qubit state out of the system faster. Yet, increasing this parameter is at odds with the requirement of having a low rate of spontaneous emission, which enables longer qubit lifetimes. Fortunately, because the readout resonator frequency is detuned from the qubit transition frequency, a filter may be engineered that suppresses spontaneous emission (at the qubit transition frequency) while being transparent at the readout resonator frequency.

There are a number of ways of realizing this so-called Purcell filter, including the use of quarter-wave stubs [18], low-Q bandpass filters [19, 20], and the crosstalk between qubit and 50 Ω environment [21]. Our design takes the form of a stepped impedance Purcell filter (SIPF) consisting of alternating sections of low- and high-impedance (of Z_{lo} and Z_{hi}, respectively) CPW, shown in Fig. 4a [22]. The SIPF consists of five sections connected by 40 μm tapers, with the low-impedance sections forming the ends. The measured (at 4.2 K) and modeled transmission and reflection characteristics of the SIPF drawn in Fig. 4b, show a passband around the readout resonator frequency at 6.4 GHz arising from the filter's resonant structure. Likewise, a large stopband from $\sim 2 - 6$ GHz, whose suppression increases with the number of sections in the SIPF and the impedance asymmetry $\alpha = Z_{\text{hi}}/Z_{\text{lo}}$, gives a suppression of 20 dB at a typical qubit frequency of 5 GHz. The fact that the SIPF may be placed on a separate die helps reduce spurious package modes [23] and opens the possibility of modular implementation.

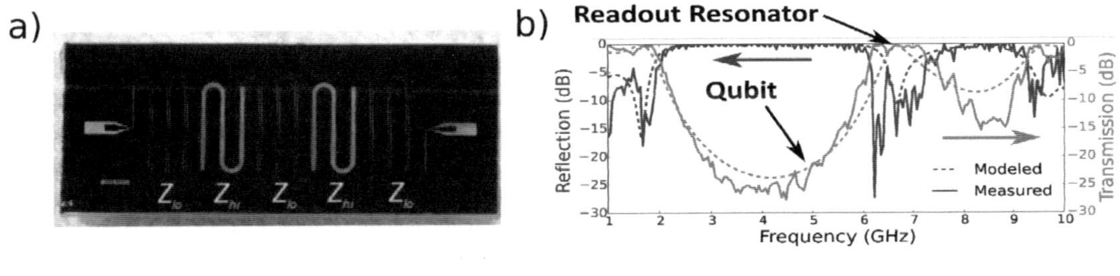

Figure 4. a) Five section SIPF consisting of alternating sections of CPW with impedance Z_{lo} and Z_{hi}. b) Modeled and measured transmission and reflection characteristics of the SIPF depict a passband around the readout resonator frequency and stopband around the qubit transition frequency. Adapted from Ref. [22].

We investigate the performance of the SIPF by studying a large-κ resonator coupled to a flux-tunable superconducting transmon device. The device and SIPF are backmounted to a printed circuit board (PCB) by a copper push block, shown in Fig. 5. The PCB routes signals through internal 50 Ω buried stripline traces, which are wire-bonded at the openings of the windows to feedlines on the qubit/resonator device and SIPF. The length of the trace connecting the device and SIPF is tailored such that the lowest mode supported by the trace lies far below the qubit transition frequency. Qubit lifetime is modeled and measured and plotted in Fig. 5b. Here, the green line indicates the maximum lifetime without a Purcell filter. The blue dots are measured qubit lifetime showing an order of magnitude improvement over the green line, and agree well with a model including the Purcell filter and an intrinsic (i.e., not readout related) loss channel corresponding to a maximum lifetime of $T_1^{\rm intr} \approx 55$ μs depicted in purple. This agrees with similar devices measured at IBM. The black curve corresponds to the maximum lifetime given only by the Purcell effect, indicating the potential for improving the qubit lifetime by another order of magnitude or two by addressing the intrinsic loss mechanisms of the qubit. Hence, the SIPF allows fast measurement without the concomitant degradation in qubit lifetime that comes from increased resonator linewidth. Further details can be found in Ref. [22].

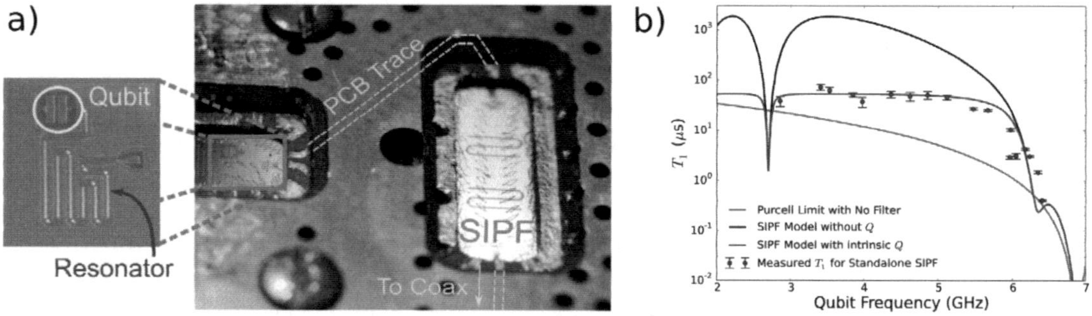

Figure 5. a) SIPF connected to a qubit device by a buried stripline 50 Ω PCB trace. b) Measured T_1 as a function of frequency of the device in a). Blue dots are measured data, which fits well with a T_1 model assuming an intrinsic $T_1^{\rm intr} \approx 55$ μs in purple. The green (black) line corresponds to the T_1 limit of the device with (without) the SIPF. The mode around 3 GHz corresponds to the 50 Ω PCB trace. Adapted from Ref. [22].

4.2. Josephson Parametric Converter

The second component in our output scheme, which enables us to perform high-fidelity, single-shot measurement of the qubit state, is a Josephson parametric amplifier, known as the Josephson parametric converter (JPC) [24]. Before we explain how the integration of this preamplifier prior to the HEMT amplifier enables us to achieve this goal, it is important to briefly review the JPC physics and mode of operation. The JPC is a nondegenerate, three-wave mixing device capable of amplifying weak microwave signals at the quantum limit. It consists of two half-wave microwave transmission line resonators, as shown in the circuit diagram of the JPC in Fig. 6a and the JPC micrograph in Fig. 6b. The two resonators intersect in a Josephson ring modulator (JRM) at their centers. The JRM consists of four Josephson junctions arranged in a Wheatstone bridge configuration, which serves as a nonlinear, dispersive medium capable of performing three-wave mixing processes. The JRM has three eigenmodes: two differential and one common. The two differential modes are supported by the corresponding fundamental resonance mode of the JPC resonators, which we refer to as the signal and idler resonators. In order to operate the device as an amplifier, a strong, coherent microwave drive, referred to as the pump, is applied to the device whose frequency is the sum of the resonance frequencies of the signal and idler resonators. The pump drive gives rise to a common mode excitation of the JRM, which leads to parametric amplification of the reflected signal and idler tones incident on the device ports. Since the JPC operates in reflection, a second circulator must be added between the circulator and isolators of Fig. 3 to route the signal to the JPC and then to the output line. In order to feed the signal, idler, and pump tones into the device, state-of-the-art JPCs employ two commercial, off-chip, broadband 180 degree hybrids as shown in Figs. 6a and 6c, which are connected to the feedlines of the signal and idler resonators. In this configuration, the signal and idler tones are fed though the Δ port of the corresponding hybrids, while the pump tone is fed through one of the Σ ports of the two hybrids (the Σ port of the other hybrid is terminated by 50 Ω). Furthermore, in order for the JRM to operate as a three-wave mixing device, a dc circulating current in the outer loop of the JRM is needed to facilitate the interaction between the three modes. Such a circulating current can be induced by applying an external magnetic field through the JRM loop. In standard JPC devices, this magnetic flux is applied by using an external magnetic coil mounted on the JPC package, as shown in Fig. 6c. It is worthwhile noting that the main purpose of the four inner Josephson junctions of the JRM, shown in Fig. 6a, is to inductively shunt each Josephson junction of the JRM (that is on the external loop). The role of this inductive shunting is to lift the hysteretic response of the JRM versus flux and make the resonance frequencies of the JPC tunable [25].

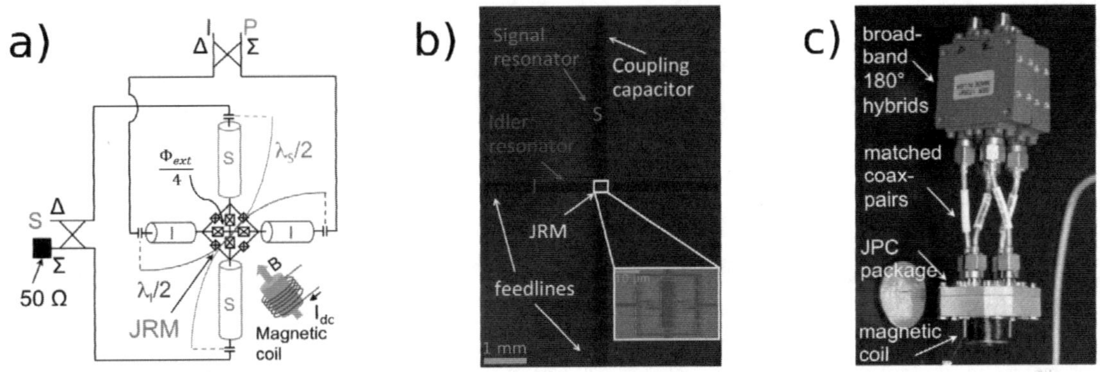

Figure 6. a) Standard JPC circuit. The JPC consists of two microwave resonators with different fundamental resonance modes, denoted signal (S) and idler (I). The two orthogonal resonators intersect in the middle at a JRM. The device circuit also includes two off-chip, broadband 180 degree hybrids and an external magnetic coil. b) Micrograph of a JPC device implemented using microstrip resonators, and c) A photo of a JPC device featuring the JPC package, the two hybrids, and the external magnetic coil.

Following this brief introduction of the JPC, we now turn to explain how the JPC allows us to achieve a high-fidelity, single-shot (i.e., without averaging over many rounds) readout of the qubit state. One simple way to explain it is using the illustration shown in Fig. 7, which qualitatively compares (not to scale) between the signal-to-noise ratio (SNR) of the output signal achieved using the "HEMT amplifier only" readout scheme versus the SNR achieved using the readout scheme which employs the "JPC as a preamplifier before the HEMT". In the illustrations of Fig. 7, we represent the output microwave readout signals and the corresponding noise in the IQ-plane, where I and Q represent the two quadratures of the microwave field (which can be directly measured by mixing down the readout signal and integrating at room temperature). In Fig. 7a, we show an example of a readout signal at the reference plane that is dominated by vacuum noise and exhibits a symmetric superposition of $|0\rangle$ (blue, left) and $|1\rangle$ (red, right) states of the qubit. The blue and red circles represent 2D histograms of the qubit state measurements, which given a suitable binning, count the number of integrated shots that occur in each bin. In this picture, the distance between the center of each 2D histogram (blue or red) and the origin (of the IQ-plane) represents the amplitude of the microwave signal, whereas the radius of the circle represents the standard deviation of the signal noise. In Fig. 7b, we show an illustration of the same readout signal as in Fig. 7a after being amplified by the HEMT and referred back to the reference plane. As can be seen in this figure, the added noise by the HEMT, which can be on the order of 20 input photons at the signal frequency, causes a very large overlap of the two 2D histograms representing the two qubit states, as the width of the 2D histograms are very large compared to the separation of the corresponding centers. As a result of this large overlap, the ability to distinguish with high-fidelity between the qubit states $|0\rangle$ and $|1\rangle$ is diminished. In contrast, by using a quantum-limited amplifier (QLA), such as the JPC, as a preamplifier before the HEMT, which adds noise on the order of a half input photon at the signal frequency, the SNR characterizing the readout signal, when referring back to the reference plane, is mainly dominated by the added noise of the preamplifier. This leads to a separation between the centers of the 2D histograms representing the qubit states, which exceeds their width. This, in turn, allows us to determine the qubit state with high fidelity using the integrated signal of a single measurement.

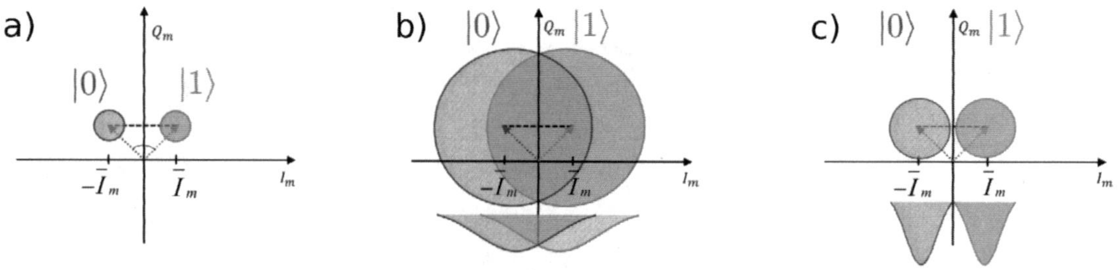

Figure 7. Illustrations that qualitatively show the 2D histograms of the signal and the corresponding noise in the IQ-plane for two different readout schemes as explained in the text (not to scale). In these illustrations, the readout signal carries information about the qubit state and the qubit is in an equal superposition of $|0\rangle$ and $|1\rangle$. a) A representation of the readout signal and the corresponding vacuum noise as seen at the reference plane (depicted in Fig. 3) before being amplified. b) A representation of the readout signal and corresponding noise in the "HEMT amplifier only" readout setup, when referred back to the reference plane. c) A representation of the readout signal and corresponding noise in the "JPC as a preamplifier before the HEMT" scheme, when referred back to the reference plane.

In Fig. 8, we demonstrate the ability to do fast, high-fidelity readout using a JPC amplifier and SIPF. The blue, red, and black curves of Fig. 8 represent double Gaussian fits to qubit-state histogram measurements, corresponding to the qubit being prepared in the $|0\rangle$, $|1\rangle$, and $|0\rangle + |1\rangle$ states, respectively. The readout measurements of the qubit are amplified by a JPC and HEMT. The achieved readout fidelity in this measurement is 96.5% with a measurement time $\tau_m = 370$ ns. Such a short measurement time is made possible due to the large readout resonator bandwidth (κ) of this device. Moreover, in this setup we retain a relatively long qubit lifetime (T_1) of 48 μs, despite the large bandwidth of the readout resonator, that is due to the presence of the SIPF.

4.3. Logic and Control Electronics
The third component needed for running quantum algorithms with active feedback is fast decision and control. Current methods of using a digitizer connected to CPU over PCIe does not allow real-time processing of data, because of the data transfer overhead and incurred software

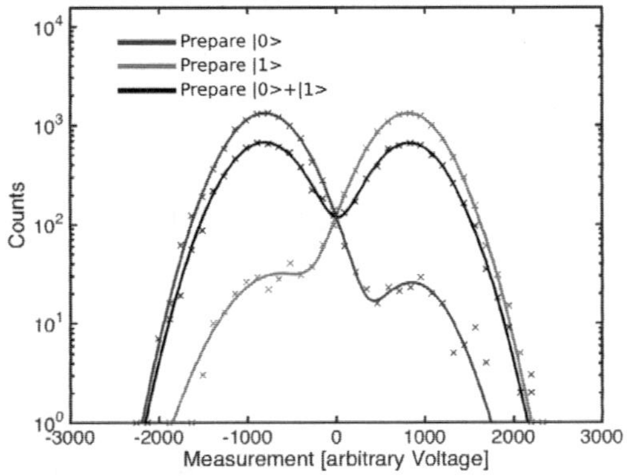

Figure 8. Readout histograms obtained by projecting 2D histogram measurements, as in Fig. 7, on the I-quadrature. The blue, red, and black curves represent double Gaussian fits to the histogram measurements of the qubit, after being prepared in the $|0\rangle$, $|1\rangle$, and $|0\rangle + |1\rangle$ states, respectively.

latency. Implementing logic directly on a Field Programmable Gate Array (FPGA), which has a deterministic latency because logic is performed in hardware within a fixed number of clock cycles, allows real time processing of data. Therefore, execution of decisions can be made based on measurement feedback with minimal latency. We implement logic on an FPGA connected to an analog-to-digital converter (ADC). The FPGA is also connected to a digital-to-analog converter (DAC) for stimulus for active control and a CPU over PCIe for testing. FPGA electronics for automated decision and control has recently become of interest to researchers in the quantum computing field because it enables low-latency signal processing in hardware [26, 27]. Our setup, consisting of modular components for testing and evaluation, is depicted in Fig. 9. Here, the IF signal is digitized and integrated using a matched filter [28], and thresholded to assign the value of $|0\rangle$ or $|1\rangle$ to the measurement. The histogram of this data is inset in Fig. 9 and indicates a readout fidelity of 95% when using both a JPC and HEMT for amplification and the SIPF. This is similar to readout fidelity using the standard digitizer. Current experiments involve active qubit reset, characterization of qubit stimulus from the DAC, and closing the loop to run the entire experiment from the FPGA. While current setups are done at room temperature, it is possible in the future that such decisions and control operations could be done at cryogenic temperatures [29].

5. Conclusion and Future of Readout Improvements

In summary, we presented three ingredients necessary to achieve fast, high-fidelity readout. A large resonator bandwidth for fast measurement without degrading the qubit lifetime is allowed by using an SIPF. Single-shot readout is facilitated by using a QLA, in our case a JPC in conjunction with a HEMT amplifier. And fast decision and control provided by a custom FPGA with ADC and DAC boards. Moreover, the FPGA can provide stimulus for active qubit reset and the large resonator bandwidth also aids in quickly clearing the cavity of readout photons.

As quantum processors become larger and contain more qubits, it becomes necessary to shrink and possibly eliminate as many microwave components as possible. For example, a new version of our JPC replaces the magnetic coil with an on-chip flux line, and uses an on-chip power divider to obviate the need for bulky hybrids [30]. Such a compact design is expected to be instrumental towards realizing directional amplifiers based on the JPC [24] and on-chip circulators [31, 32]. Current experiments at IBM involve one input and one output channel per

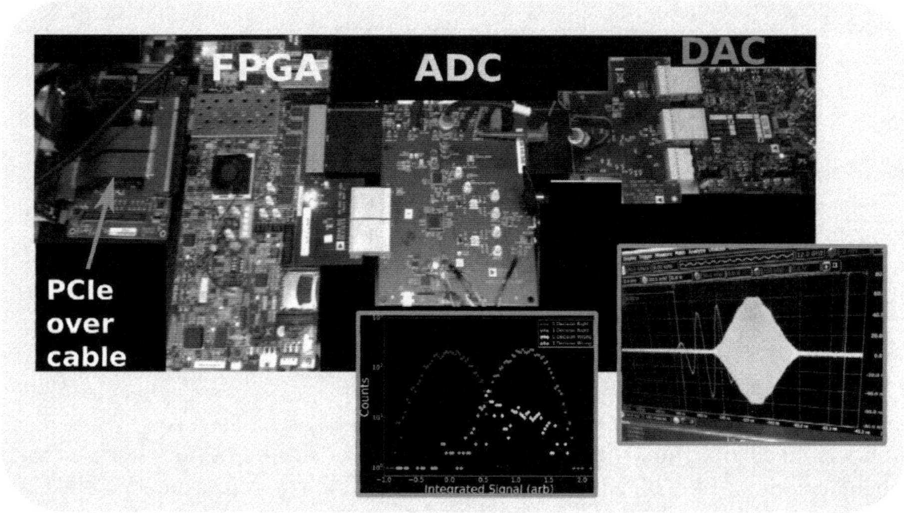

Figure 9. Custom control electronics. An FPGA is connected to a control computer (for testing) via a high-throughput PCIe over cable. Mounted on the FPGA is a ADC that integrates the readout signal using pre-calculated kernel weights of a matched filter and a DAC that can be used for stimulus, i.e. active qubit reset.

qubit, and schemes for adding more qubits per channel are being investigated. Furthermore, our group is exploring frequency multiplexing schemes, which would require the use of a wide-band QLA, such as a traveling wave parametric amplifier [33, 34], channel multiplexing, and fast microwave switching [35].

Acknowledgments

We thank M. Brink, M. B. Rothwell, and G. A. Keefe for fabrication of qubit devices and J. R. Rozen, J. Rohrs, and Y.-K.-K. Fung for experimental contributions. Work regarding the SIPF was done in conjunction with Y. Liu and A. A. Houck. We acknowledge support from IARPA under contract W911NF-10-1-0324.

References

[1] Blais A, Huang R S, Wallraff A, Girvin S M and Schoelkopf R J 2004 *Physical Review A* **69** ISSN 1050-2947
[2] Wallraff A, Schuster D I, Blais A, Frunzio L, Majer J, Kumar S, Girvin S M, Schoelkopf R J and Huang R S 2004 *Nature* **431** 162 ISSN 1476-4687
[3] Shor P W 1997 *SIAM Journal on Computing* **26** 1484–1509 ISSN 0097-5397
[4] Grover L K 1997 *Physical Review Letters* **79** 325 ISSN 0031-9007
[5] Yung M H, Whitfield J D, Boixo S, Tempel D G and Aspuru-Guzik A 2014 Introduction to quantum algorithms for physics and chemistry *Advances in Chemical Physics* ed Kais S (Hoboken, New Jersey: Jon Wiley & Sons) pp 67–106
[6] Shor P 1994 *Proceedings 35th Annual Symposium on Foundations of Computer Science* 124–134 ISSN 0272-5428
[7] Chow J M, Gambetta J M, Corcoles A D, Merkel S T, Smolin J A, Rigetti C, Poletto S, Keefe G A, Rothwell M B, Rozen J, Ketchen M B and Steffen M 2012 *Physical Review Letters* **109** 060501 ISSN 0031-9007
[8] Barends R, Kelly J, Megrant A, Veitia A, Sank D, Jeffrey E, White T C, Mutus J, Fowler A G, Campbell B, Chen Y, Chen Z, Chiaro B, Dunsworth A, Neill C, O'Malley P J J, Roushan P, Vainsencher A, Wenner J, Korotkov A N, Cleland A N and Martinis J M 2014 *Nature* **508** 500 ISSN 0028-0836
[9] Harty T P, Allcock D T C, Ballance C J, Guidoni L, Janacek H A, Linke N M M, Stacey D N and Lucas D M 2014 *Physical Review Letters* **113** 2–6 ISSN 10797114
[10] Nigg D, Müller M, Martinez E A, Schindler P, Hennrich M, Monz T, Martin-Delgado M A, Blatt R, Muller

M, Martinez E A, Schindler P, Hennrich M, Monz T, Martin-Delgado M A and Blatt R 2014 *Science* **345** 302–305 ISSN 1095-9203
[11] Koch J, Yu T M, Gambetta J M, Houck A A, Schuster D I, Majer J, Blais A, Devoret M H, Girvin S M and Schoelkopf R J 2007 *Physical Review A* **76** 042319 ISSN 1050-2947
[12] Paraoanu G S 2006 *Physical Review B* **74** 140504 ISSN 1098-0121
[13] Rigetti C and Devoret M H 2010 *Physical Review B* **81** 134507 ISSN 1098-0121
[14] Chow J M, Corcoles A D, Gambetta J M, Rigetti C, Johnson B R, Smolin J A, Rozen J R, Keefe G A, Rothwell M B, Ketchen M B and Steffen M 2011 *Physical Review Letters* **107** 080502 ISSN 0031-9007
[15] Zhou X Q, Kalasuwan P, Ralph T C and O'Brien J L 2013 *Nature Photonics* **7** 223 ISSN 1749-4885
[16] Elitzur A C and Vaidman L 1993 *Foundations of Physics* **23** 987–997 ISSN 0015-9018
[17] Kwiat P, Weinfurter H, Herzog T, Zeilinger A and Kasevich M 1995 *Annals of the New York Academy of Sciences* **755** 383–393 ISSN 00778923
[18] Reed M D, Johnson B R, Houck A A, DiCarlo L, Chow J M, Schuster D I, Frunzio L and Schoelkopf R J 2010 *Applied Physics Letters* **96** 203110 ISSN 00036951
[19] Jeffrey E, Sank D, Mutus J Y, White T C, Kelly J, Barends R, Chen Y, Chen Z, Chiaro B, Dunsworth A, Megrant A, O'Malley P J J, Neill C, Roushan P, Vainsencher A, Wenner J, Cleland A N and Martinis J M 2014 *Physical Review Letters* **112** 190504 ISSN 0031-9007
[20] Kelly J, Barends R, Fowler A G, Megrant A, Jeffrey E, White T C, Sank D, Mutus J Y, Campbell B, Chen Y, Chen Z, Chiaro B, Dunsworth A, Hoi I C, Neill C, O'Malley P J J, Quintana C, Roushan P, Vainsencher A, Wenner J, Cleland A N and Martinis J M 2015 *Nature* **519** 66 ISSN 0028-0836
[21] Bronn N T, Magesan E, Masluk N A, Chow J M, Gambetta J M and Steffen M 2015 *IEEE Transactions on Applied Superconductivity* **25** 1700410 ISSN 1051-8223
[22] Bronn N T, Liu Y, Hertzberg J B, Corcoles A D, Houck A A, Gambetta J M and Chow J M 2015 *Applied Physics Letters* **107** 172601 ISSN 0003-6951
[23] Srinivasan S J, Corcoles A D, Magesan E, Bronn N T, Hertzberg J, Gambetta J M, Steffen M and Chow J M 2015 *Bulletin of the American Physical Society* **60** BAPS.2015.MAR.J39.4
[24] Abdo B, Sliwa K, Frunzio L and Devoret M 2013 *Physical Review X* **3** 031001 ISSN 2160-3308
[25] Roch N, Flurin E, Nguyen F, Morfin P, Campagne-Ibarcq P, Devoret M H and Huard B 2012 *Physical Review Letters* **108** 147701 ISSN 0031-9007
[26] Ofek N, Petrenko A, Heeres R, Reinhold P, Leghtas Z, Vlastakis B, Liu Y, Frunzio L, Girvin S M, Jiang L, Mirrahimi M, Devoret M H and Schoelkopf R J 2016 *Nature* **536** 441 ISSN 0028-0836
[27] Bultink C C, Rol M A, O'Brien T E, Fu X, Dikken B C S, Dickel C, Vermeulen R F L, de Sterke J C, Bruno A, Schouten R N and DiCarlo L 2016 *Physical Review Applied* **6** 034008 ISSN 2331-7019
[28] Ryan C A, Johnson B R, Gambetta J M, Chow J M, da Silva M P, Dial O E and Ohki T A 2015 *Physical Review A* **91** 022118 ISSN 1050-2947
[29] Colless J I and Reilly D J 2012 *Review of Scientific Instruments* **83** 023902 ISSN 00346748
[30] Abdo B, Chavez-Garcia J M, Brink M, Keefe G and Chow J M 2016 (*Preprint* arXiv:1609.00401)
[31] Kerckhoff J, Lalumière K, Chapman B J, Blais A and Lehnert K K 2015 *Physical Review Applied* **4** 034002 ISSN 2331-7019
[32] Sliwa K M, Hatridge M, Narla A, Shankar S, Frunzio L, Schoelkopf R J and Devoret M H 2015 *Physical Review X* **5** 041020 ISSN 2160-3308
[33] Macklin C, O'Brien K, Hover D, Schwartz M E, Bolkhovsky V, Zhang X, Oliver W D and Siddiqi I 2015 *Science* ISSN 0036-8075
[34] Vissers M R, Erickson R P, Ku H S, Vale L, Wu X, Hilton G C and Pappas D P 2016 *Applied Physics Letters* **108** 012601 ISSN 0003-6951
[35] Pechal M, Besse J C, Mondal M, Oppliger M, Gasparinetti S and Wallraff A 2016 (*Preprint* arXiv:1606.01031)

IOP Conf. Series: Journal of Physics: Conf. Series **834** (2017) 012004 doi:10.1088/1742-6596/834/1/012004

Development of cryogenic CMOS Readout ASICs for the Point-Contact HPGe Detectors for Dark Matter Search and Neutrino Experiments

Zhi Deng[1,2], Li He[3], Feng Liu[1,2], Yinong Liu[1,2], Tao Xue[1,2], Yulan Li[1,2], Qian Yue[1,2]

[1] Department of Engineering Physics, Tsinghua University, Beijing, 100084, China
[2] Key Laboratory of Paticle & Radiation Imaging, Ministry of Education, Beijing, 100084, China
[3] Joint Research Center, Nuctech Company Limited, Beijing, 100084, China

dengz@mail.tsinghua.edu.cn

Abstract. The paper presents the developments of two cryogenic readout ASICs for the point-contact HPGe detectors for dark matter search and neutrino experiments. Extremely low noise readout electronics were demanded and the capability of working at cryogenic temperatures may bring great advantages. The first ASIC was a monolithic CMOS charge sensitive preamplifier with its noise optimized for ~1 pF input capacitance. The second ASIC was a waveform recorder based on switched capacitor array. These two ASICs were fabricated in CMOS 350 nm and 180 nm processes respectively. The prototype chips were tested and showed promising results. Both ASICs worked well at low temperature. The preamplifier had achieved ENC of 10.3 electrons with 0.7 pF input capacitance and the SCA chip could run at 9 bit effective resolution and 25 MSPS sampling rate.

1. Introduction

The point-contact high purity germanium detector (PC-HPGe) technology has been successfully demonstrated for dark matter search and neutrino-less double beta decay experiments [1-4]. Benefit from the low capacitance (~1 pF) of the small point contact, the equivalent noise charge (ENC) for PC-HPGe can be lowered down to several electrons. Hence low threshold and high energy resolution can be achieved, which are of great importance to the low energy limit and sensitivity for rare event detection.

However extremely low noise readout electronics are demanded. Conventionally JFETs (Junction Field Effect Transistors) are used due to their lower flicker noise [5,6]. As the detector capacitance decreases down to ~1 pF the contribution from the flicker noise becomes insignificant. In addition JFETs cannot be mounted very close to the detectors since their optimum working temperature is usually ~120 K. Hence the parasitic capacitance of the interconnection cannot be minimized. On the contrary CMOS (Complementary Metal Oxide Semiconductor) transistors can work at 77 K and can be more easily custom designed with highly integration. Several low noise CMOS preamplifier ASICs (Application Specific Integrated Circuits) have been reported both for silicon drift detectors with very small capacitance of 10-100 fF [7,8] and for HPGe detectors with detector capacitance of ~10 pF [9,10]. The noise of the preamplifiers for silicon drift detectors could be very low (<10 e) but they were not optimized for ~1 pF detector capacitance. Our previously designed CMOS preamplifier has achieved energy resolution of 512 eV FWHM at 59.5 keV with a 3 pF PC-HPGe detector [11]. It turned out to be very promising and a newly developed ASIC dedicated for smaller capacitance and its test results will be present in this paper.

On the other hand, pulse shape analysis is essential for HPGe detector signal processing [12] and the waveform of the preamplifier output needs to be sampled and stored for further digital signal processing. In current dark matter and neutrino experiments the preamplifier outputs were sent out from the cryostat through 10-meter-long cables and the waveforms were sampled and digitized by room-temperature fast ADCs. In order to reduce the number of the cables as well as cryogenic feedthroughs, a multi-channel SCA (Switched Capacitor Array) ASIC was proposed here to record the signal waveforms of multiple detectors in an analogue way and then to send them out in serial by multiplexing. The SCA structure was adopted due to its simplicity, which in principle could easily work at 77K. A SCA ASIC named CASCA [13] originally developed for gas detectors at room temperature was used for feasibility study. The details of the circuit design and the test results under cryogenic environment will be described in the following sections.

2. Circuit design

2.1. The CMOS charge sensitive preamplifier ASIC

The charge sensitive preamplifier consisted of a core amplifier, providing effective gain and bandwidth, a feedback capacitor and the reset circuits, as shown in Fig.1. The core amplifier has a transconductance amplifier stage and an output stage, which is able to drive ~1 m long cable. The feedback capacitor was 50 fF. A reset switch was used since it had almost zero noise contribution when it was open. The pulse reset signal was generated either by the on-chip window discriminators or externally. A calibration capacitor of 50 fF was also integrated. The chip was implemented in a CMOS 350 nm process. All the circuit simulation was done using the transistor model provided by the manufacturer, which was not valid for such low temperature down to 77 K. Hence the bias currents of all circuit branches can be adjusted through an external resistor.

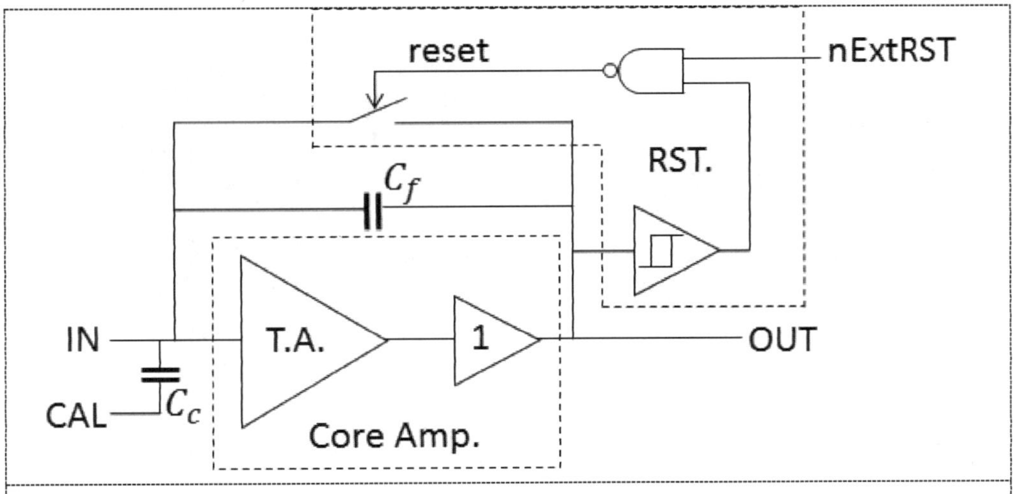

Figure 1. The architecture of the charge sensitive preamplifier.

2.1.1. The core amplifier. A traditional folded cascode amplifier structure was adopted for the transconductance amplifier, as shown in Fig.2. A PMOS input transistor was chosen for its lower flicker noise coefficient. The size of the transistor (600 μm/0.35 μm) was optimized for 1 pF input capacitance and its bias current was 0.8 mA at 77 K. The output stage was designed to drive large negative signals without slew rate limit [14]. The total consumed current of the core amplifier was 2.8 mA.

IOP Conf. Series: Journal of Physics: Conf. Series **834** (2017) 012004 doi:10.1088/1742-6596/834/1/012004

2.1.2. The reset circuits. The minimum sized NMOS and PMOS transistors were used in parallel for the reset switches. Dummy transistors were evenly placed to minimize the charge injection. A resistor was also put in series to the switch transistors to increase the stability of the amplifier and hence to supress the oscillation during the reset. Two discriminators were used to set the upper and lower thresholds respectively. A pulse reset signal with ~100 ns fixed duration was asserted automatically when the output of the preamplifier was out of the range defined by the two thresholds, which can be adjusted externally.

2.1.3. The chip layout and the test PCB. The layout of the preamplifier is shown in Fig.3 and the size of the chip is 2.5 mm x 2.3 mm. A test PCB was designed with the die wire bonded on it. The test PCB was mounted on the cold finger of the cryostat, as shown in Fig.4.

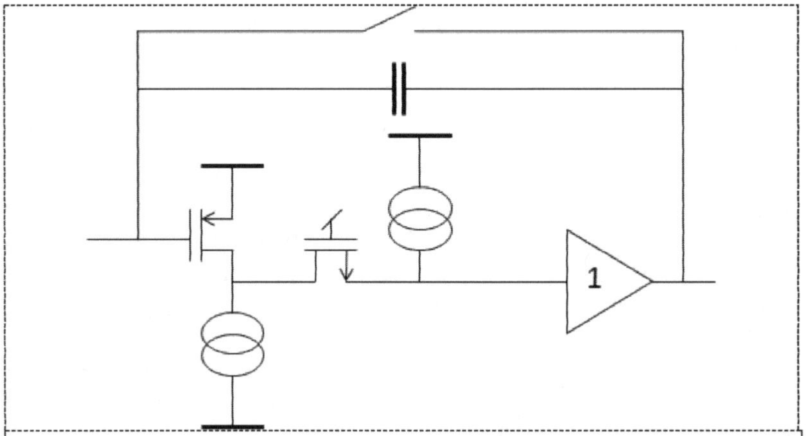

Figure 2. The schematics of the folded cascode amplifier.

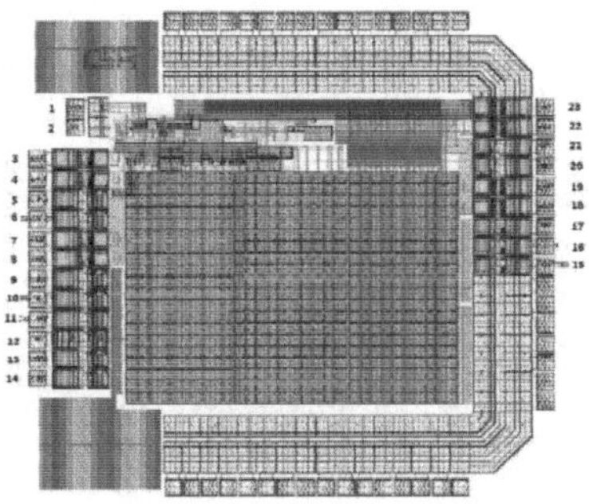

Figure 3. The layout of the preamplifier, in size of 2.5 mm x 2.3 mm.

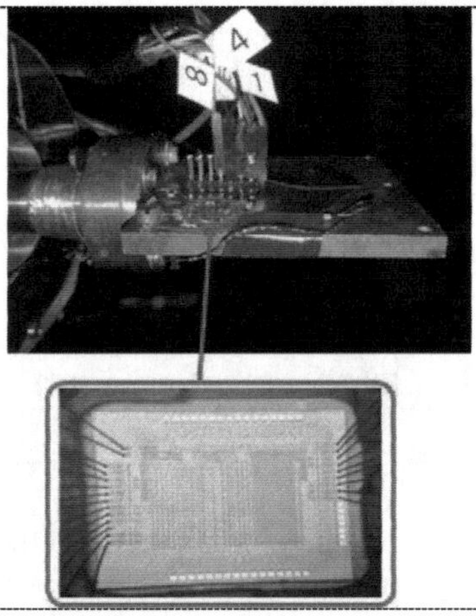

Figure 4. The test PCB mounted on the cold finger of the cryostat.

2.2. The waveform sampling ASIC based on switched capacitor array

The simplified block diagram of the SCA ASIC was shown in Fig.5. It consisted of 32 channels of switched capacitor array, a readout amplifier for each channel, a 32:1 multiplexer and a differential output driver. The control logic was also integrated. The chip was fabricated in a CMOS 180 nm process and was originally developed for time projection chamber gas detectors. Only a short description of working principle and circuits will be introduced below and the detailed circuit design can be found in Ref.13.

During the sampling phase, the write switch *WR* of each capacitor was turned on for one clock period in sequence and the input voltage would be stored on the corresponding capacitor. If there was no signal detected, the input signal was continuously sampled on the capacitor cells in a cycling way. There were 64 capacitor cells in total for each channel. Once a trigger asserted, the sampled voltage on the capacitors would be frozen and waited for readout. The readout switch *RD* could be selected by the 6 bit address bus. During the readout phase, the selected sampling capacitor was connected as the feedback capacitor of the readout amplifier. The voltage stored was then present at the output of the amplifier, which was sent out channel by channel by multiplexing and was converted to differential current signals. The chip was designed for working at room temperature but it might work at 77 K due to that there were no complex amplifiers and all the bias currents could be externally adjusted.

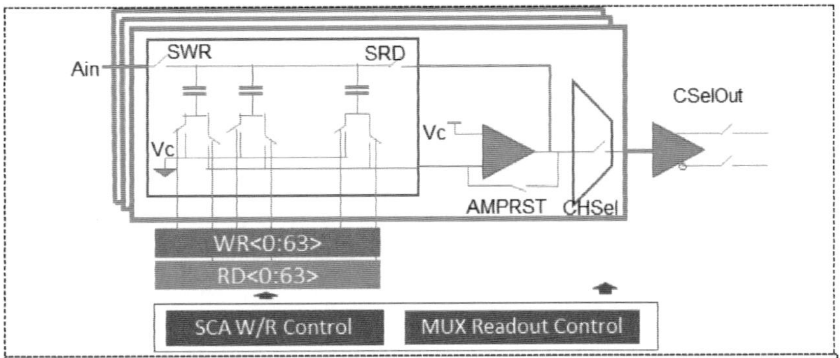

Figure 5. The simplified block diagram of the SCA ASIC

3. Test results

The performance of these two ASICs under cryogenic temperature will be described below.

3.1. *The charge sensitive preamplifier*

The external bias resistor was set to 520 Ω and its bias current I_{bias} was measured at different temperatures, as shown in Fig.6. The bias currents of all circuit branches in the core amplifier were proportional to the master bias I_{bias}. The master bias current had a significant change from 480 μA at 300 K to 135 μA at 100 K and it was estimate to be about 100 μA at 77 K, which was very close to the simulation results.

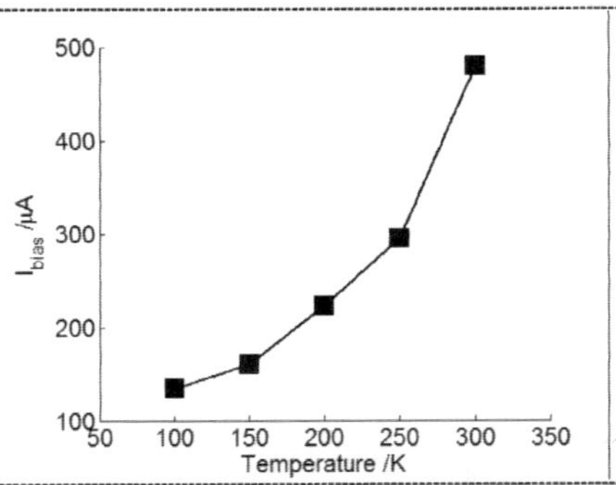

Figure 6. The temperature dependence of the bias current.

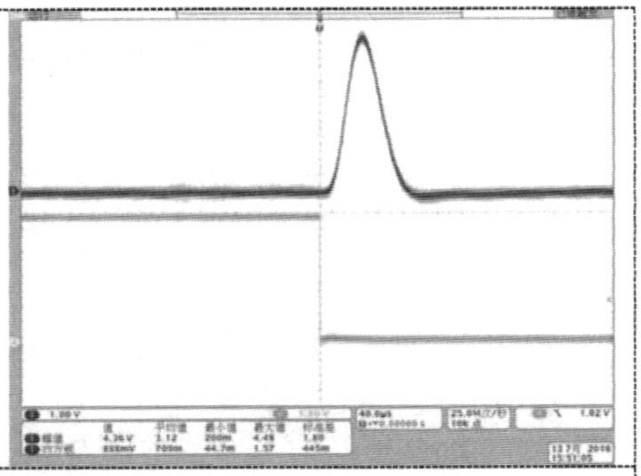

Figure 7. The transient signals. Pink: input step pulse; Blue: shaper output

IOP Conf. Series: Journal of Physics: Conf. Series **834** (2017) 012004 doi:10.1088/1742-6596/834/1/012004

The preamplifier was tested by injecting a step pulse through a calibration capacitor, mimicking the detector signal. The injected signal charge can be calculated by $C_{cal} \cdot \Delta V$, where C_{cal} is the calibration capacitance and ΔV is the amplitude of the step pulse. The output of the preamplifier was sent to a commercial spectroscopy amplifier (Canberra 2022) for shaping and noise filtering. The transient signal waveforms were shown in Fig.7. The charge gain $G = {V_{out}}/{Q_{in}}$ of the amplifiers was then derived. The r.m.s standard deviation of the output baseline σ_{vo} was measured by the oscilloscope and the ENC was calculated by ${\sigma_{vo}}/{G}$. ENC was dependent on the input capacitance and the shaping time and the curves of ENC vs. input capacitance at different shaping time at 300 K and 100 K were shown in Fig.8. The bias currents at different temperature were set to about the same (~100 μA). Due to the test setup the temperatures for the noise measurement at 100 K were actually in the range of 90-100 K. The minimum of 10.3 electrons was achieved with 0.7 pF input capacitance at 12 μs shaping time at 100 K.

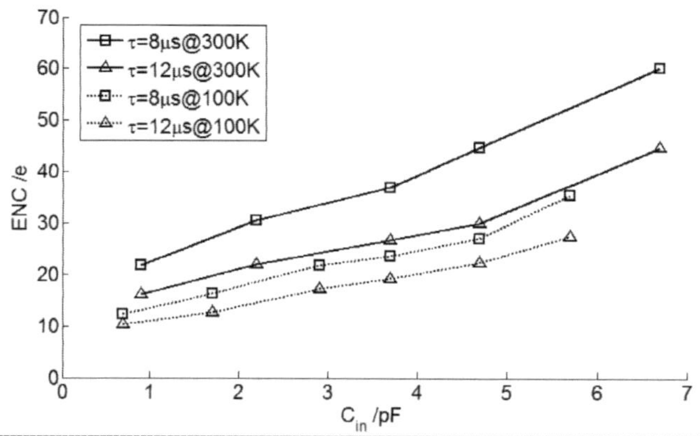

Figure 8. ENC vs. input capacitance

3.2. The SCA

The test setup for the SCA ASIC was shown in Fig.9. The SCA ASIC was put on the cold finger of the cryostat vacuum chamber. The power supply, the output and the control signals including clock and trigger were connected out through ~1 m long cables. The control signals were generated by a FPGA evaluation board and the analog output was digitized by an ADC board. The data was acquired through the USB interface to the computer.

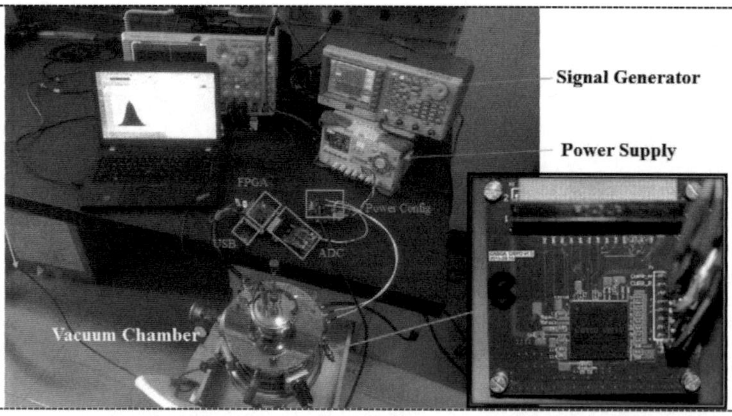

Figure 9. The test setup for the SCA ASIC

IOP Conf. Series: Journal of Physics: Conf. Series **834** (2017) 012004 doi:10.1088/1742-6596/834/1/012004

A standard dynamic ADC test procedure was conducted to evaluation the SCA waveform sampling performance. A sine signal with 200 kHz frequency and almost full amplitude (~1 V) was sampled and readout by the SCA ASIC. The collected samples were fitted with the sinusoidal function (Fig.10) and then the resolution was estimated from the standard deviation of the residual. An example of residual distribution was shown in Fig.11.

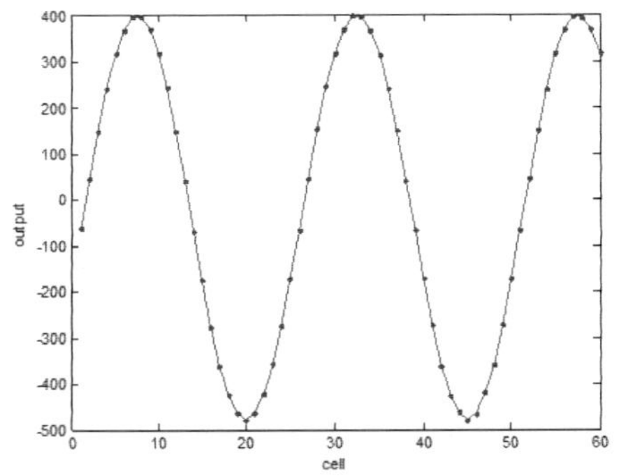

Figure 10. Fitting the readout SCA samples with sine waveform.

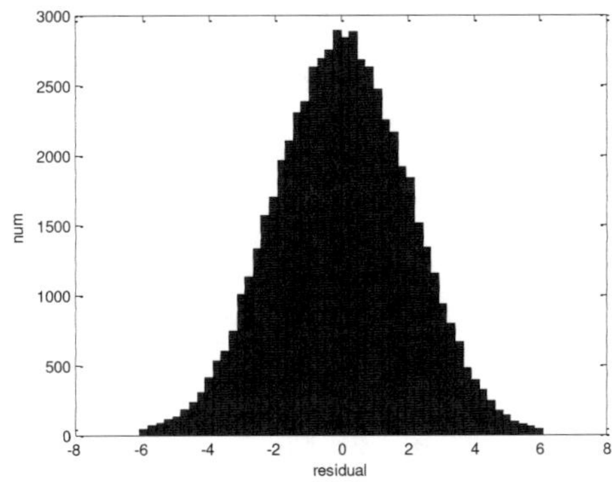

Figure 11. The distribution of the residual.

The dynamic tests were done with sampling frequency up to 25 MSPS at 300 K and 120 K respectively, which was limited by the test setup. In order to adjust the bias current to be the same (~50 μA), the external resistor had to be changed from 1.4 kΩ at 300 K to 118.2 Ω at 120 K. The r.m.s. standard deviations were measured to be 2.2 mV and 1.9 mV at 300 K and 120 K, corresponding to 8.8 bit and 9 bit effective resolution.

4. Summary

Two ASICs have been developed for PC-HPGe detectors for dark matter search and neutrino experiments. Both ASICs worked well at cryogenic temperature. The CMOS charge sensitive preamplifier had achieved ENC of 10.3 electrons with 0.7 pF input capacitance and the SCA ASIC could run at 9 bit effective resolution and 25 MSPS sampling rate. All these prototype chips cannot fully satisfy the requirements for the experiments. More tests and ASIC developments are undergoing for better noise and resolution performances in the near future.

Our experience with CMOS 350 nm and 180 nm processes showed that simple cascode and operational amplifiers could work easily at 77 K with external bias strategy. However it may need more accurate low temperature transistor model for further optimization.

Acknowledgments

This work was partially supported by the Natural Science Foundation of China under grant no. 11375100. The authors would also like to thank ICC (Shanghai, China) for providing the MPW service.

References

[1] Zhao W, Yue Q, Kang K J, et al. 2013 *Phys. Rev. D* **88(5)** 052004
[2] Aalseth C E, Barbeau P S, Colaresi J. 2013 *Phys. Rev. D* **88(1)** 012002
[3] Agostini M, Allardt M, Andreotti E, et al. 2013 *Phys. Rev. Lett.* **111(12)** 122503
[4] Abgrall N, Aguayo E, Avignone F T, et al. 2014 *Adv. in High Energy Phys.* **2014** 365432
[5] Riboldi S, Cattadori C, D'Andragora A, et al. 2010 *IEEE Nucl. Sci. Symp. & Med. Imag. Conf.* p 1386

[6] Barton P, Luke P, Amman M, et al. 2011 *IEEE Nucl. Sci. Symp. & Med. Imag. Conf.* p 1976
[7] De Geronimo G, Rehak P, Ackley K, et al. 2010 *IEEE Trans. on Nucl. Sci.* **57(3)** 1654
[8] Bombelli L, Fiorini C, Frizzi T, et al.2011 *IEEE Nucl. Sci. Symp. & Med. Imag. Conf.* p 1972
[9] Riboldi S, Pullia A, Zocca F, et al. 2007 *IEEE Nucl. Sci. Symp. & Med. Imag. Conf.* p 353
[10] Pullia A, Zocca F, Riboldi S, et al. 2010 *IEEE Trans. on Nucl. Sci.* **57(2)** 737
[11] Zhu X, Deng Z, Li Y, et al. 2011 *IEEE Nucl. Sci. Symp. & Med. Imag. Conf.* p 766
[12] Barbeau P S, Collar J I, Tench O. 2007 *J. Cosmol. Astropart. Phys.* **2007(09)** 009
[13] Zhang H Y, Deng Z, He L, et al. 2015 *IEEE Nucl. Sci. Symp. & Med. Imag. Conf.* p 1
[14] Pullia A, Zocca F. 2006 *IEEE Nucl. Sci. Symp. & Med. Imag. Conf.* p 354

IOP Conf. Series: Journal of Physics: Conf. Series **834** (2017) 012005 doi:10.1088/1742-6596/834/1/012005

First-ever test and characterization of the AMS standard bulk 0.35 μm CMOS technology at sub-kelvin temperatures

A Rhouni, O Gevin, X de la Broïse, JL Sauvageot, V Revéret, L Rodriguez

CEA-Saclay, DRF, IRFU, F91191, Gif/Yvette Cedex, France

amine.rhouni@cea.fr or olivier.gevin@cea.fr

Abstract. From medical imaging to particle physics passing, among others, by space applications, integrated readout electronics (ICs) in CMOS technologies are often adopted. When a high sensitivity and a low noise level are required, cooling of detectors and readout electronics is the recommended solution. To maintain a constant cooling temperature, they very often operate at nitrogen and helium-4 liquids temperatures, respectively 77 K and 4.2 K. At these temperatures, Spice parameters of MOSFET transistors may be found in the literature. However, their performances at sub-kelvin temperatures remain unknown because of a lack in scientific publications thereupon. CEA Astrophysics division's focal plane arrays-based bolometers are cooled at 0.1 K. The front-end electronics also. However, a CMOS technology was characterized for the first time at sub-kelvin temperatures. It is shown by measured n and p channel transistors' I-V that the AMS 0.35 μm standard bulk CMOS technology, is still performing at 0.1 K. Despite some specific effects on silicon behaviour at cryogenic temperatures, performances are very satisfactory.

1. Introduction

In a continuing effort to improve the sensitivity and reliability of bolometers arrays for the far-infrared space astronomy, the Astrophysics and the Electronics, Detectors and computing divisions are associated to the CEA/Leti institute in a join development.

The next generation of bolometers arrays will be cooled below 0.1 K to improve the electric and thermal responses and to reduce the intrinsic noise of the pixels. All this elements will contribute to achieve the required sensitivity for future Space Projects such as CMB B-mode exploration or future FIR observatories like SPICA, a common ESA-JAXA Project [1]. This sensitivity is expressed as the noise equivalent power (NEP). The impedance of resistive silicon bolometers increases as the temperature decreases following Efros-Shklovskii model [2]. In our design, it may reach few GΩ or more at 0.1 K, which enables to design focal plane arrays with a NEP down to some atto-Watts/√Hz in the low frequency band of the detectors response. However, due to this huge bolometers impedance, minimizing the input capacitance becomes necessary to preserve the detector bandwidth. This is achievable by a very close integration of the front-end electronics. The large number of pixels, in the required focal planes, implies also an interconnection scheme significantly more complex than simple routing tracks. Leti has developed through its contribution to the ESA Herschel/PACS instrument launched in 2009 [3], a technological know-how of silicon bolometers arrays hybridized by indium bonds on a cooled CMOS front-end electronics working at 0.3 K. The next generation is derived from this mature technology, by growing the detectors directly on the MOS circuit. Figure 1 shows the above IC principle and in

IOP Conf. Series: Journal of Physics: Conf. Series **834** (2017) 012005 doi:10.1088/1742-6596/834/1/012005

particular, the electrical interconnections made by metal sinks (copper nails of 3 µm in diameter). This implies an identical operating temperature for both detectors and readout circuits.

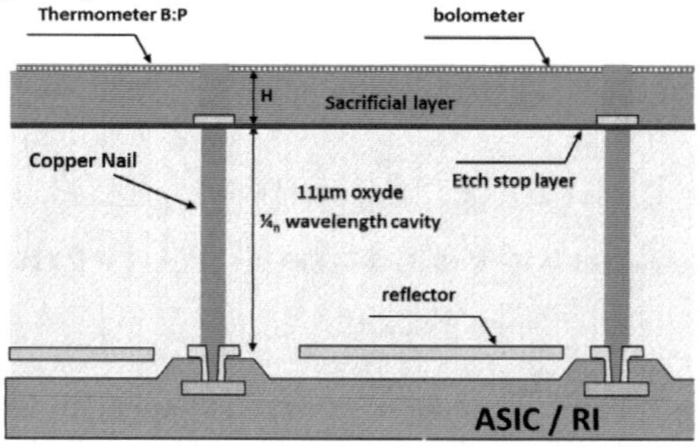

Figure 1. CEA/Leti « Above IC » process: bolometers arrays are manufactured on the ASIC wafer.

To characterize the chosen CMOS technology at 0.1 K, an Application Specific Integrated Circuit (ASIC) and a test printed board were designed and fabricated. In this paper, we first describe the developed electronics system and the test setup. Then, we discuss the I-V measurement results and observed effects.

2. Designed ASIC for technology characterization

An ASIC was designed to characterize the AMS 0.35 µm CMOS technology elementary components at 0.1 K. It hosts MOSFET transistors with various gate sizes and different values of poly silicon resistors and capacitors as shown in figure 2. Transistors are grouped into two blocs: small channel length (0.35 to 0.75 µm) and large channel length (10 µm). They have common gate, source and n-well substrate. This allows reducing the number of analog pads and integrating 23 transistors, 5 resistors and 3 capacitors on the same chip (2.3 mm x 2.3 mm). When a transistor is tested, drains of parallel transistors are connected to their common source to avoid current leakages. Gate-to-source and gate-to-drain capacitors do not influence I-V measurements (DC measurements). Nonetheless, for AC analysis, the total induced capacitance on the common gate should be considered because of its effect on the frequency bandwidth.

The ASIC is packaged in a JLCC-68 and gold bonding wires are preferred for thermal and electrical conductivities purpose. On the package cavity, a calibrated thermometer is bonded to continually probe the ASIC's temperature (Figure 3).

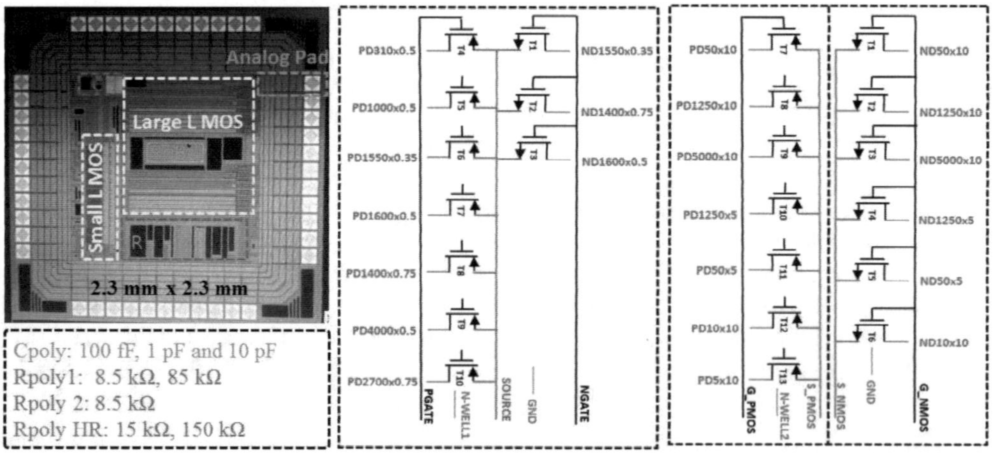

Figure 2. The ASIC microphotography and the two blocks of transistors (small and large L).

3. Cryogenic test setup

A 3He/4He-cryocooler is used to cool the ASIC. The dilution of Helium-3 in the Helium-4 allows to achieve temperatures close to absolute zero. Three days are required to cool down and one day to warm up. Hence, it is important to take benefit of each cooling by performing a maximum number of tests. Thus, a PCB was designed to allow testing all ASIC components through a 21-μD connector (see figure 3). Two cooling operations are required to perform the 23 transistors I-V.

Thermal loads and power dissipation are an issue in cryogenic applications. Tested subsystems self-heating determines the coldest achievable temperature by the cryocooler. A PCB adds extra materials such as ceramic, plastic, copper … to the refrigerator. Their thermal resistance limits the ability to cool the sample down to 0.1 K. To help cooling the ASIC, a copper bench vise was made and mounted on the coldest plate of the cryocooler (see figure 3 (a)). It applies a strong mechanical/thermal contact on the package from both, the top side and the bottom side through a rectangular hole in the PCB. The thermometer on the cavity measures 0.1 K, when the coldest plate is at 0.02 K. If a temperature below 0.1 K is sought, the thermal link between the copper and the package's ceramic must be improved (e.g. by using thermal grease or anisotropic layers). Here, the goal is 0.1 K. The interconnection interface between the μD connector and the ASIC pads is made by 24 jumpers. 470 μF capacitors are used for 1Hz-filtering in noise analysis, which are not discussed in this paper. All mounted components (jumpers, capacitors, resistors, μD connector, thermometer, wires), were cryogenically qualified within previous works.

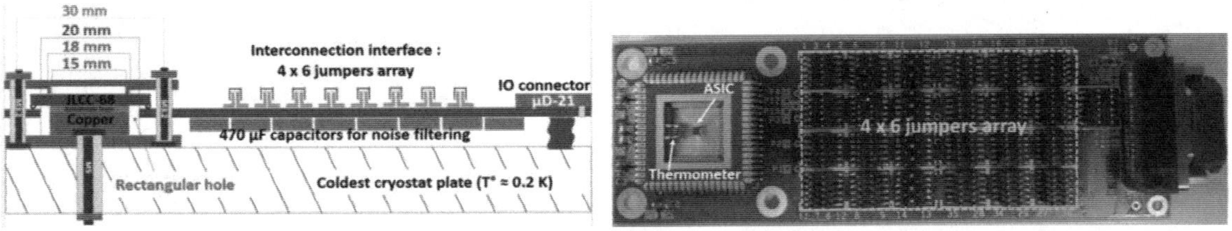

Figure 3. (a) Conceptual drawing of the PCB and the cryogenic test setup. **(b)** Photo of the manufactured 8 layers PCB for the ASIC test at 0.1 K.

4. I-V measurements results

A parameter analyzer Keithley SCS-4200 was used to perform transistors I-V as sketched in figure 4. In this section, output and transfer characteristics are discussed for: 3 PMOS transistors (10 μm x 10 μm, 50 μm x 10 μm and 1250 μm x 10 μm) and 1 NMOS transistor (10 μm x 10 μm). MOSFETs with large gates were primarily tested because of the need thereafter to design a low flicker noise front-end electronics. Plotted PMOS I-V data are shown in figure 5.

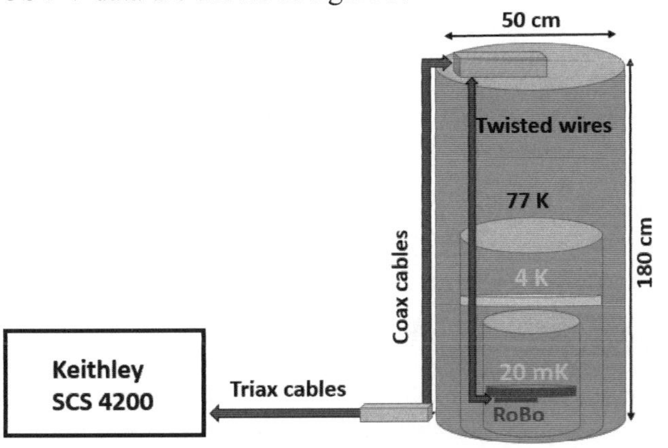

Figure 4. Cryogenic test setup

For future foreseen applications, the maximum power dissipation of the front-end electronics was estimated. Considering a 3 years mission lifetime and a focal plane made of 10000 bolometers, the total power budget available for the frond-end readout ASIC is 2 µW. By using time domain multiplexing, the available bias current per pixel is in the order of few tens of nA when the bias voltage is 100 or 200 mV. Considering this power budget and the limited cooling power of the cryocooler, characteristics were performed for low drain currents (maximum of 150 nA), when the drain-to-source voltage is 500 mV. The n-well substrate was always set at source voltage of 3 V.
Prima facie, transfer and output characteristics look as usual for PMOS transistors. The drain current is still depending on the gate/drain voltage and on the gate size. Outputs characteristics on figure 5 (b), (c) and show weak hysteresis for small PMOS, which disappear for the largest gate 1250 µm x 10 µm.
Plots (a) of figure 5 show that at a constant bias current, the transconductance gm is still depending on the transistor gate size and figure 6 (a) shows the evolution of gm versus the gate size for two drain currents. The 10 µm x 10 µm PMOS transistor's measured gm at 0.1 K is plotted in figure 6 (b) in function of the drain current and compared to both, the sub-threshold and above-threshold gm slopes (respectively 1 and 0.5). The transistor operates between the two regimes.
From these first characteristics and despite the hysteresis, it could be deduced that PMOS transistors perform well at 0.1 K.

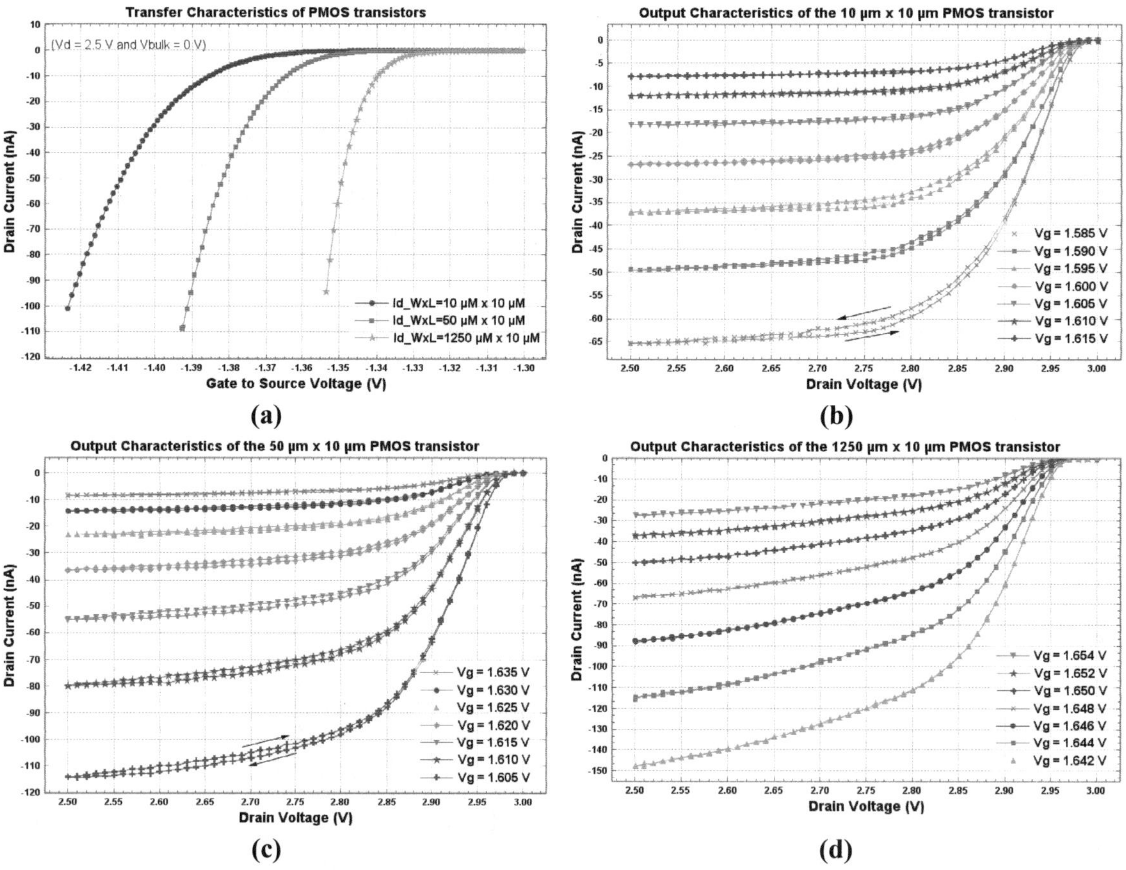

Figure 5. Measured PMOS transfer and output characteristics at 0.1 K.

NMOS transistors are more affected at sub-kelvin temperatures compared to PMOS. Two phenomena are met at 0.1 K (figure 7): kink effect and drain current overshoot. Both depend on the sweep delay (i.e. holding time before each measure) and the sweep direction.

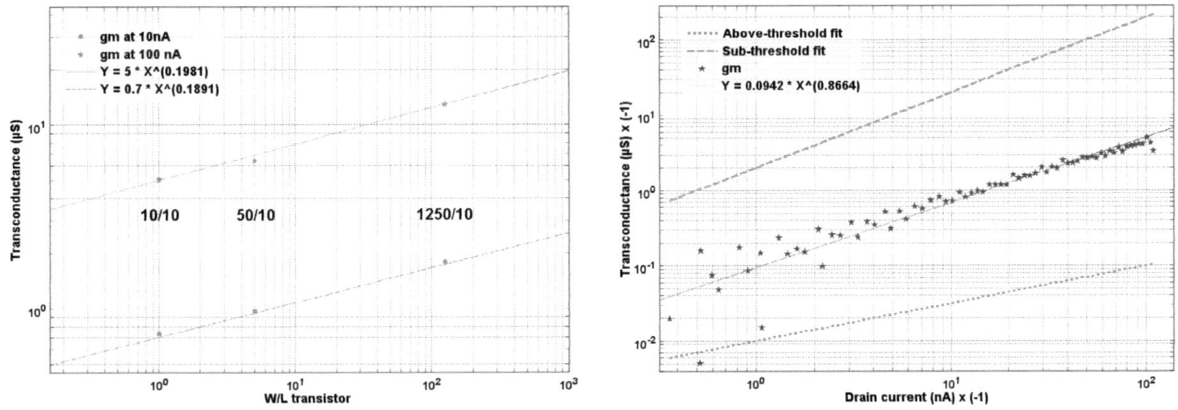

(a) **(b)**

Figure 6. (a) PMOS transistors gm versus the gate size for two drain currents at 0.1 K **(b)** The 10 µm x 10 µm PMOS gm versus the drain current. Measurements are performed at 0.1 K.

The current overshoot is observed when sweeping from low-to-high drain voltage as shown in plots (a) of figure 7. It has a maximum amplitude (1.65 µA) for a sweep delay of 250 ms, which consistently decreases until disappearing when the sweep delay increases. Plots (b) of figure 7 show the kink effect (i.e. low increase of the slope $\Delta I/\Delta V$ in the voltage range 0.1 V - 0.9 V and strong increase at 0.1 V). This is usually observed on SOI CMOS technologies at Helium-4 temperature because of the silicon substrate's freezing out [4]. It seems that it also depends on the sweep delay: it decreases when the sweep delay increases.

The relationship between both phenomena and the sweep delay suggests a kind of capacitive effect in NMOS transistors, which is not the case for PMOS transistors. This would be explained by the difference between substrates. Indeed, a PMOS transistor is made on N-doped substrate (N-Well), which seems unaffected by the freezing out.

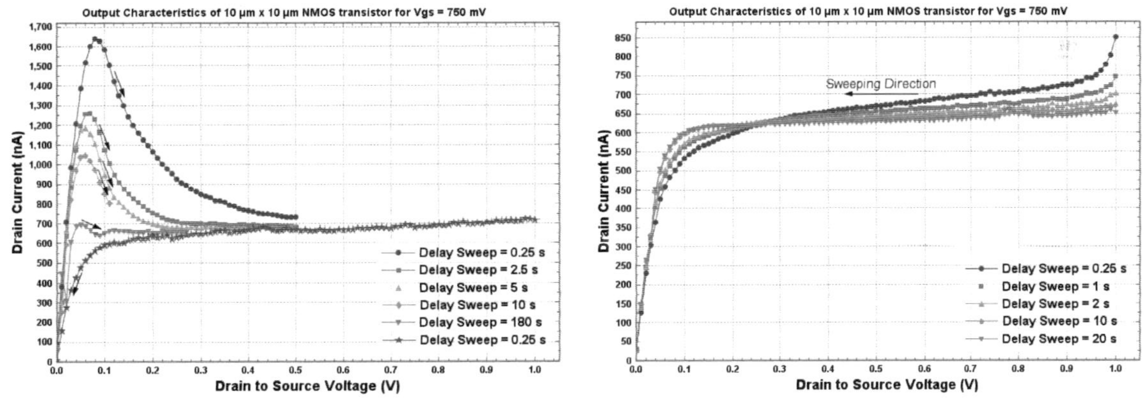

(a) **(b)**

Figure 7. 10 µm x 10 µm NMOS output characteristics. (a) Show the drain current overshoot and (b) the kink effect.

5. Conclusion

It is shown in this paper that PMOS and NMOS transistors are still performing at 0.1 K for low drain currents. PMOS transistors have usual characteristics and could be used to design various electronic functions. NMOS transistors seem to be affected by the freeze out of the silicon substrate. However, they could be used to design digital functions or switches. For analog functions, the operating point must be chosen beside the kink effect and the current overshoot areas.

I-V characteristics of small channel length transistors will be performed to expand applications range and design possibilities. Also, low frequency noise analysis are under study to provide transistors noise figures at different sub-kelvin temperatures for different gate sizes. This will allow predicting flicker and thermal noises levels regardless of the temperature.

Finally, by leveraging all of these measurements results, the 10000 bolometers focal plane arrays readout electronics will be designed and manufactured in the AMS standard bulk 0.35 μm CMOS technology.

References

[1] T. Nakagawa, et al, « The next-generation infrared astronomy mission SPICA under the new framework », Proc. SPIE 9143, Space Telescopes and Instrumentation 2014, Optical, Infrared and Millimeter Wave, 91431, 2014.

[2] Efros, A. L., and B. I. Shklovskii. "Coulomb gap and low temperature conductivity of disordered systems." Journal of Physics C: Solid State Physics 8.4 (1975): L49.

[3] Poglitsch, Albrecht, et al. "The photodetector array camera and spectrometer (PACS) on the Herschel space observatory." Astronomy & astrophysics 518 (2010): L2.

[4] Mercha, Abdelkarim, et al. " Linear kink effect" induced by electron valence band tunneling in ultrathin gate oxide bulk and SOI MOSFETs." IEEE Transactions on Electron Devices 50.7 (2003): 1675-1682.

Acknowledgments

Authors wish to acknowledge the European Research Council for its support through the ORISTARS program and would like to express, to Bernadette DOMENGES from LAMIPS, their appreciation for her collaboration and her technical support on the ASIC.

Also, the authors would like to thank Keithley-Tektronik France for their commercial offer.

Fabricating with crystalline Si to improve superconducting detector performance

A D Beyer[1], M I Hollister[2], J Sayers[2], C F Frez[1], P K Day[1], and S R Golwala[2]

[1]Jet Propulsion Laboratory, California Institute of Technology, Pasadena, CA
[2] California Institute of Technology, Pasadena, CA

Andrew.d.beyer@jpl.nasa.gov

Abstract. We built and measured radio-frequency (RF) loss tangent, tan δ, evaluation structures using float-zone quality silicon-on-insulator (SOI) wafers with 5 μm thick device layers. Superconducting Nb components were fabricated on both sides of the SOI Si device layer. Our main goals were to develop a robust fabrication for using crystalline Si (c-Si) dielectric layers with superconducting Nb components in a wafer bonding process and to confirm that tan δ with c-Si dielectric layers was reduced at RF frequencies compared to devices fabricated with amorphous dielectrics, such as SiO_2 and Si_xN_y, where tan δ ~ 10^{-3}. Our primary test structure used a Nb coplanar waveguide (CPW) readout structure capacitively coupled to LC resonators, where the capacitors were defined as parallel-plate capacitors on both sides of a c-Si device layer using a wafer bonding process with benzocyclobutene (BCB) wafer bonding adhesive. Our control experiment, to determine the intrinsic tan δ in the SOI device layer without wafer bonding, also used Nb CPW readout coupled to LC resonators; however, the parallel-plate capacitors were fabricated on both sides of the Si device layer using a deep reactive ion etch (DRIE) to access the c-Si underside through the buried oxide and handle Si layers in the SOI wafers. We found that our wafer bonded devices demonstrated $F\cdot\delta = (8 \pm 2) \times 10^{-5}$, where F is the filling fraction of two-level states (TLS). For the control experiment, $F\cdot\delta = (2.0 \pm 0.6) \times 10^{-5}$, and we discuss what may be degrading the performance in the wafer bonded devices as compared to the control devices.

1. Introduction

Future space-borne missions using spectrometers operating in the sub-millimeter (sub-mm) and millimeter wave (mm-wave) region require technological advancements over the state-of-the art to meet flight mission design goals and objectives. Superconducting kinetic inductance detectors (KIDs), superconducting microstripline structures in phased-array antennae, and superconducting microstripline networks for filtering and light routing hold great promise for space-mission instrument needs if lower loss dielectrics can be utilized in construction. Currently, amorphous dielectrics such as SiO_2 and Si_xN_y are used due to their simplicity in fabrication. The loss tangent, tan δ, in amorphous dielectrics is approximately 10^{-3}. Amorphous, hydrogenated Si (a-Si:H) has been reported to have lower loss with tan δ between 0.3-5 × 10^{-5}[1]. However, among common semiconductor fabrication materials, crystalline Si (c-Si) has been reported to have the best reported tan δ, with tan δ as low as 10^{-6} [2]. We briefly describe the benefits which c-Si may afford in superconducting KIDs and microstriplines before describing how we pursued c-Si in this report.

Superconducting phased-array antennae rely on microstriplines formed from superconductor-dielectric-superconductor structures to define beam shapes from the sky and route signals through on-chip filters before detection by other superconducting detectors such as transition edge sensors (TESs) [3] or KIDs [4]. Losses in the microstrip require the detectors to be placed near the antenna, potentially leading to unwanted absorption of light that has not been coupled through the antennae. Furthermore, the efficiency of the phased array antenna is ultimately affected by loss in the microstrip. By reducing this loss through through the use of c-Si dielectrics, as opposed to amorphous SiO_2 or Si_xN_y, detectors can both be moved further away from the antennae to reduce stray pickup and larger, multiscale antennae covering more spectral bandwidth could be designed.

Spectrometers-on-a-chip, such as SuperSpec [5], also use superconducting microstripline networks to effectively couple light onto the chip and route the light to different detectors to produce lower mass, smaller spectrometers. Currently, the resolving power of such spectrometers is R $\sim\lambda/\Delta\lambda \sim 1/\tan\delta$=1,000. By utilizing c-Si, R may improve to on the order of 100,000. Such a development would enable resolved extragalactic mm/submm spectroscopy, where intrinsic line widths are $\Delta\nu/\nu \sim 10^{-4}$ to 10^{-3}.

Finally, KID noise performance degrades when amorphous dielectrics with two-level states (TLS) reside in the portion of the detectors where electric fields exist [6]. Experiments reveal that moving KIDs to other substrates, such as sapphire that lack TLS, greatly improves baseline noise performance. Due to the crystalline structure, c-Si will not contribute TLS, thus expanding the range of applications were KIDs can obtain photon noise-limited performance [7].

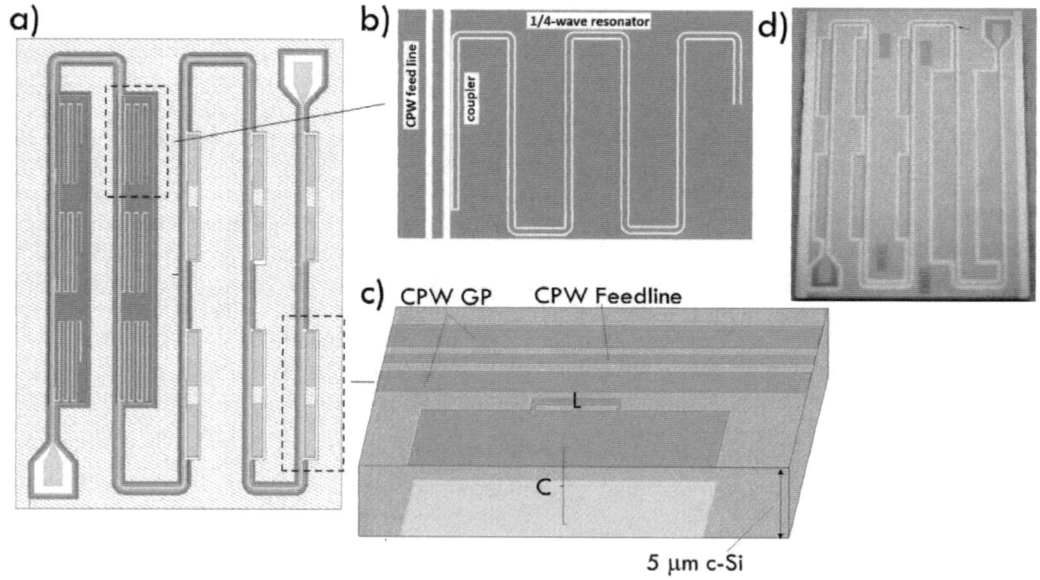

Figure 1: Our wafer bonded RF test structure. a) A schematic of the whole device, which consisted of six LC resonators and six ¼-wave resonators made using superconducting Nb and crystalline Si (c-Si) of 5 μm thickness. The Nb was patterned on both sides using a robust wafer bonding fabrication process. We determine the loss tangent, tan δ, by measuring the frequency shift of the LC resonators versus temperature. There are two bond pads for connection to the coplanar waveguide (CPW) readout in the bottom left and top right of the device. [b)The ¼-wave resonators serve as metal loss test structures but are not examined in this work.] c) A schematic slab-cut of the LC resonators showing the CPW feedline readout capacitively coupled to a length of Nb serving as the inductor (L). The capacitor for each of the LC resonators is a parallel-plate capacitor (C). d) An optical picture of the actual device.

IOP Conf. Series: Journal of Physics: Conf. Series **834** (2017) 012006 doi:10.1088/1742-6596/834/1/012006

Here, we report on our effort to use this extremely low loss material for dielectric layers in superconducting KIDs and microstriplines. The scope of this report focuses on validating low loss at RF frequencies, and future work will explore sub-mm and mm-wave validation of low loss tangent in c-Si. We quantify the loss tangent of c-Si from two experiments using c-Si layers in silicon-on-insulator (SOI) wafers. One experiment, our control experiment, measured LC resonators at RF frequencies formed using the c-Si device layer in a SOI wafer using a deep reactive ion etch (DRIE) process to make parallel-plate superconducting Nb capacitors. Our main experiment used wafer bonding to build more robust LC resonator structures using the c-Si layer. The wafer bonded structure was robust because the parallel-plate capacitors were supported on the underside, whereas the control structure resulted in free standing capacitors that were essentially membranes of thickness equal to the c-Si thickness of 5 μm. Robust construction is important for the future production of superconducting KIDs or microstriplines to be used in actual instruments, which must survive qualifications such as shake testing.

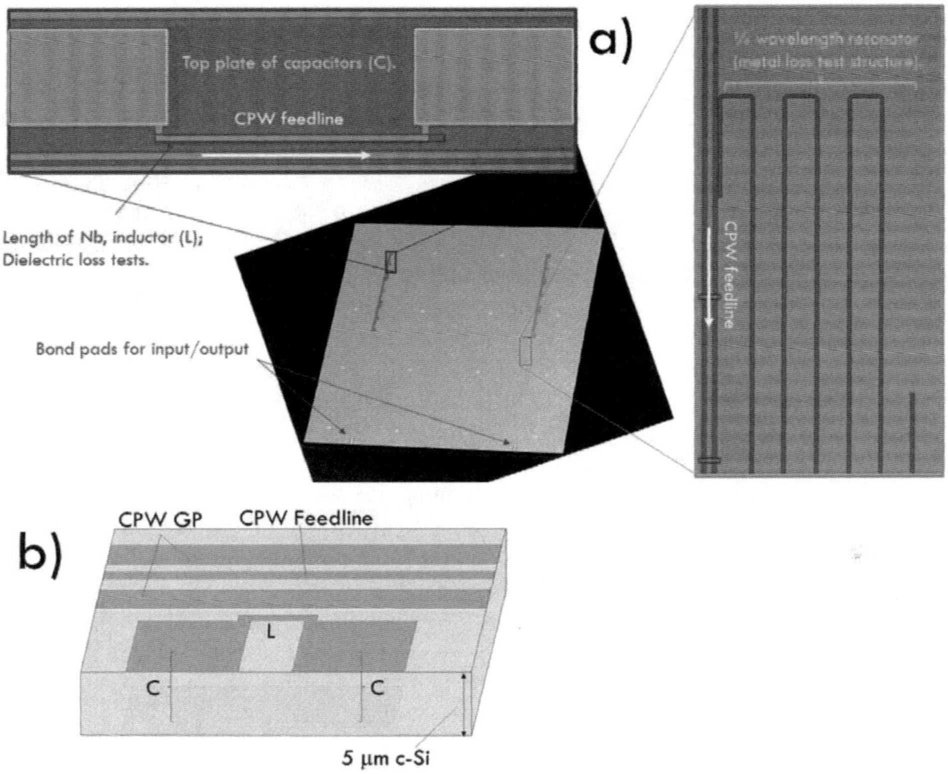

Figure 2: Our control experiment RF test structure. a) Optical images of the actual device showing zoomed in images of the LC resonators, which have 2 parallel-plate capacitors (C) and an inductor (L) formed from a length of superconducting Nb. The devices were also made with superconducting Nb and c-Si of 5 μm thickness. In contrast to the wafer bonded test device, we patterned on both sides of the c-Si using a deep reactive ion etch (DRIE) process. The DRIE results in suspended membranes of 5 μm thickness, which is not robust for larger scale production of devices or flight hardware. However, this device allows us to determine if wafer bonding affects the loss tangent of c-Si dielectrics. (Again, the metal loss test structures are shown but not discussed in this work.) b) A schematic of the LC resonators with two parallel-plate capacitors to clarify how this device differs in LC resonators from the wafer bonded device in Fig. 1. (There is no qualitative difference in using 1 or 2 parallel-plate capacitors, and the use of two in this device was based solely on the availability of photomasks.)

2. Design and fabrication

The wafer bonded RF test device and RF control test device both began with a SOI substrate consisting of a 5 μm thick, float-zone quality, c-Si device layer with > 10,000 Ω·cm sheet resistivity. Below the device layer was a buried oxide (BOX) layer of thickness 750 nm, below which was a Si handle wafer of 500 μm thickness. Both sides of the wafer were polished. The 5 μm c-Si device layer served as the c-Si for our measurements, and it was important that the c-Si was float-zone quality and with high sheet resistivity to avoid unwanted TLS states and other loss mechanisms seen in lower quality c-Si due to dopants and impurities. The superconducting metal used for these tests was superconducting Nb, with transition temperature of around 9 K.

2.1. Wafer bonded RF test structure

The main test structure we built was an RF frequency loss test structure using a wafer bonding process. We expect the wafer bonding process to produce robust fabricated structures that could be modified and infused into devices usable by instruments, especially flight projects. In Fig. 1, we show a schematic and picture of the wafer bonded RF test structure. The readout is carried out through a Nb coplanar waveguide (CPW) line that meanders from one bond pad to another, capacitively coupled to 6 LC resonators and 6 metal loss test structures. The metal loss test structures are designed for future measurements and not discussed further here. The LC resonators were designed to operate at RF frequencies between 2-4 GHz, and consisted of an inductor with a parallel-plate capacitor, as shown schematically in Fig. 1.

Initially, the SOI wafer was dipped in buffered oxide etch (BOE) for 45 seconds, to remove native oxide, and immediately loaded into our Nb sputtering chamber. A 190 nm thick Nb film was sputter deposited and subsequently etched to define a ground plane and the bottom plate for capacitors in the LC resonators. A rim of Au was then lifted off around the rim of the ground plane, to allow us to later etch down from the other side of the c-Si layer to short the top and bottom grounds together—the Nb etch, a F-based reactive ion etch (RIE) etches Si but not Au. Due to the thickness of the c-Si, we were not able to view fabrication alignment marks through the c-Si from the other side, so our final step on this side of the c-Si layer was to use DRIE to etch alignment marks through the c-Si to allow alignment from the other side. Benzocyclobutene (BCB) [8] was then applied to the patterned side of this wafer, as well as a new float-zone quality, >10,000 Ω·cm, c-Si substrate, and we bonded the two together at 10 kN of force at 250 C for 1 hour, similar to Ref. [7] and Ref. [9].

The bonded wafer stack was then sent out for commercial grinding of the Si handle of the SOI down from 500 μm to tens of μm. The remaining Si in the handle was removed using a DRIE process, and the BOX was removed using BOE. Immediately, a 115 nm thick layer of Nb was sputtered on and etched back to form the CPW ground plane, CPW feedline, and the top plates and inductors of the LC resonators. Additionally, the metal loss test structures were etched in this step. SiO_2 straps were then lifted off periodically along the length of the CPW feedline, and Nb straps were lifted off on top of the SiO_2 straps to short the ground plane together at regular intervals along the length of the CPW readout. A trench was then etched between the ground planes on the exposed side of the c-Si down to the Au rim on the other side using a DRIE process. Gold was then electroplated between the two to short the two ground planes together.

2.2. Control RF test structure

In order to determine if wafer bonding degraded the loss tangent in the c-Si layer, we also built an RF test structure using the SOI wafer and using a DRIE process to access the other side of the c-Si layer. This structure is not expected to be as robust as the wafer bonded structure because it results in membranes with thickness equal to the c-Si device layer thickness of 5 μm. The best loss tangent, tan $\delta = 10^{-6}$, was obtained from aluminium LC resonators made in a similar manner to the process we describe below [2]. Our control experiment device is shown in Fig. 2. It also had six LC resonators and six metal loss test structures, with the LC resonators consisting of an inductor and two capacitors in series.

A BOE dip, to remove native oxide, was performed before sputtering 190 nm of Nb immediately. The ground plane, straps between the ground plane along the length of the CPW readout, and the inductor and two top plates of the parallel-plate capacitors were then patterned by etching the Nb. A 230 nm layer of SiO_2 was sputtered on top of the Nb, and another 300 nm of Nb was sputtered for the CPW feedline. The SiO_2 above the LC resonators and metal loss test structures was then removed to eliminate TLS from affecting fringing electric fields in the capacitors. Finally, the back plate for the two capacitors in the LC resonator circuit was fabricated by using a DRIE and lift-off pattern technique. The Si handle of the SOI wafer was etched through using a DRIE process, the BOX layer was removed using a hydrofluoric (HF) acid vapor phase etch, and the HF etch was followed by immediate sputtering of 300 nm of Nb and lift off to avoid allowing native oxide states under the Nb.

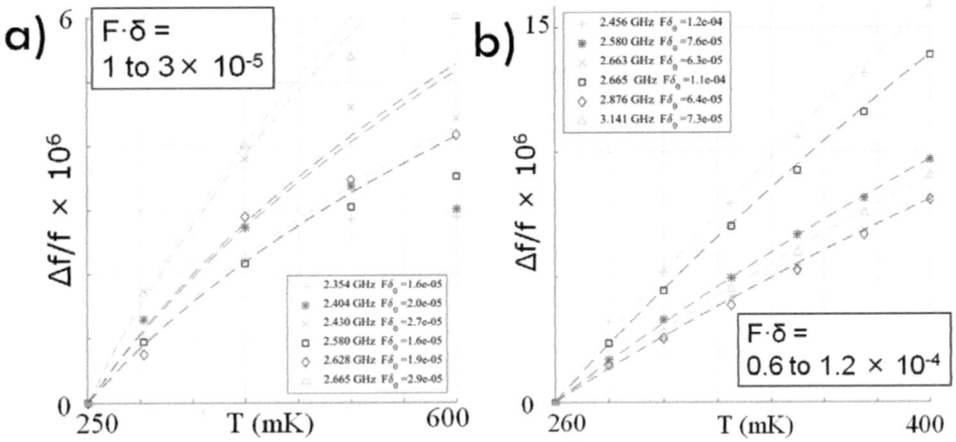

Figure 3: Normalized frequency shift vs. temperature data. a) The normalized frequency shift vs. T for the control experiment, RF test device. Points are data, and the lines are fits using Eq. (1). We find $F \cdot \delta = (2.0 \pm 0.6) \times 10^{-5}$, where F is the filling fraction of two level system (TLS) states and δ is the loss tangent. b) The normalized frequency shift vs. T for the robust, wafer bonded RF test device. Points are data, and the lines are fits using Eq. (1). We find $F \cdot \delta = (8 \pm 2) \times 10^{-5}$.

3. Measurement and analysis

To evaluate the loss tangent of our devices, we cooled the devices down to temperatures between 250 mK and 600 mK, and the resonance frequency and internal quality factor, Q_i, were characterized. We use the relationship [6] governing frequency shift versus temperature to determine the value of $F \cdot \delta$, where F is TLS filling fraction:

$$\frac{f(T)-f(0)}{f(0)} = \frac{F\delta}{\pi}\left[Re\Psi\left(\frac{1}{2} - \frac{\hbar\omega}{2j\pi k_B T}\right) - log\frac{\hbar\omega}{2\pi k_B T}\right]. \qquad (1)$$

Here, j is the imaginary constant, f(T) is the resonant frequency at temperature, T, and $\omega = 2\pi \cdot f(0)$. The values of $F \cdot \delta$ found from this analysis are nearly independent of power and reflect the low power values relevant to mm-wave and sub-mm wave power expected in the LC resonators.

IOP Conf. Series: Journal of Physics: Conf. Series **834** (2017) 012006 doi:10.1088/1742-6596/834/1/012006

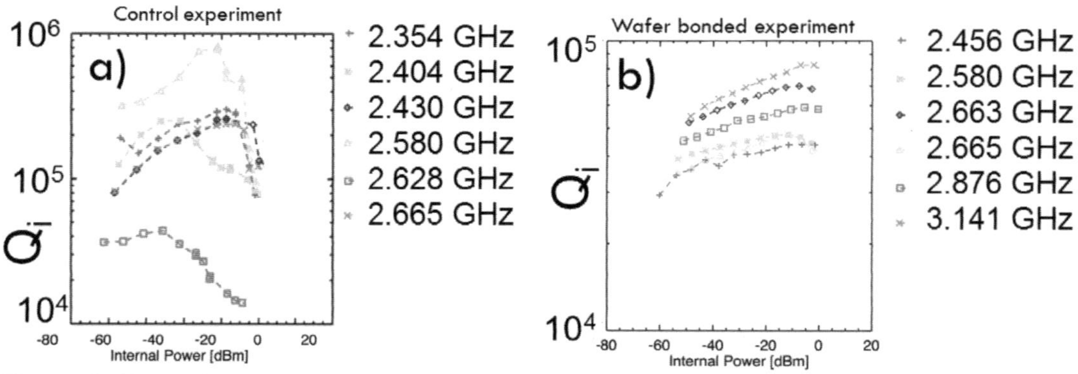

Figure 4: Internal quality factor, Q_i, versus internal power applied to readout the resonators. a) Q_i vs internal power for the control experiment, RF test device, which was fabricated with a DRIE process instead of a wafer bonding process. The Q_i values are consistently above 10^5 for almost all the resonators. b) Q_i vs internal power for the robust wafer bonded RF test device. The Q_i values are consistently above 10^4 for all the resonators. In both cases, we expect that the higher power regime (-40 dB < internal power < 0 dB) is more relevant for improvement of capacitors in kinetic inductance detectors (KIDs) where readout power can be increased to saturate TLS states [10]. For low power (internal power < -40 dB), this regime reflects the power expected to be on the resonators when illuminated by mm-wave and sub-mm wave light based on the pW power levels expected. We expect this regime applies more to improving microstripline architecture like phased array antennae and spectrometers-on-a-chip.

In Fig. 3a, the normalized frequency shift, $\Delta f/f=[f(T)-f(0)]/f(0)$, versus T is shown for the control experiment and may be fit to Eq. (1) to obtain $F\cdot\delta = (2.0 \pm 0.6) \times 10^{-5}$, which is the best loss tangent among the two RF test structures measured. In Fig. 3b, $\Delta f/f$ versus T and fits to Eq. (1) for the wafer bonded device are shown with resulting loss calculated to be $F\cdot\ \delta = (8 \pm 2) \times 10^{-5}$, about a factor of four higher in loss than the control experiment. Both measurements shown in Fig. 3 were taken with -20 dB of internal power. In Fig. 4a and 4b, we show the measured internal quality factor, Q_i, as a function of internal power for the control experiment (Fig. 4a) and with the wafer bonded device (Fig. 4b). As expected the higher loss in the wafer bonded device corresponds to lower Q_i in this device than in the control experiment.

The power dependence of Q_i is interesting when divided into two regimes, higher internal power (-40 dB < P <0) and lower internal power (P < -40 dB). The higher internal power regime is more relevant for capacitors in KIDs, because we can apply higher readout power to operate in a more favorable, higher Q_i regime where TLS are saturated. In such a regime, we obtain estimated loss tangent ($=1/Q_i$) of 1-5 $\times 10^{-5}$ in the wafer bonded test device and 0.13-1.3 $\times 10^{-6}$ in the control test device, significantly lower than that obtained from the frequency shift data because the high readout power saturates the TLS loss [10]. In contrast, the lower power regime resembles what we expect for mm and sub-mm wave operation. Such a comparison has been confirmed by comparing mm-wave and RF test devices in the MUSIC instrument, which is a microwave-KID instrument operating at mm-wave with additional diagnostic structures at RF. Indeed the pW optical power levels expected at mm-wave and sub-mm wave should correspond to low internal power. The loss tangents obtained from $\Delta f/f$ versus T measurements and fitting to Eq. (1) correspond to the values of tan δ expected at low power, even when measured at higher internal power, because the frequency shift induced by TLS is nearly independent of power and reflects the value at low power levels [10].

Both devices exhibit lower tan δ than in sputtered SiO_2 and Si_xN_y, which is typically in the range of 10^{-3}. In Ref. [2], which closely resembles our control experiment, the low power tan δ reached four times lower than what we measured, at tan $\delta = 5 \times 10^{-6}$. It is possible that our SOI wafers are of different quality than in those experiments. Another possibility is that we are affected by oxides of Nb, while

Ref. [2] used Al and had less or different thicknesses of oxide affecting performance. The wafer bonded devices are a factor of 4 worse than our control experiment. Here, the BCB layer may be close enough to the parallel-plate capacitors for the fringe electric fields to travel through the BCB and add loss and filling fraction of TLS in the wafer bonded device. In future work, we could confirm this idea by additional patterning steps using a DRIE to etch through the backside and remove the BCB from behind the resonators for proof of our conjecture. At higher power, Ref. [2] obtained tan $\delta = 10^{-6}$, and our control experiment is again a factor of a few worse while our wafer bonded device is at least a factor of 10 lower in Q_i. The same reasons of different SOI quality or oxides, as in the low power regime, would apply here too, but we continue to try to deduce other reasons for reduced performance as well. Ultimately, our wafer bonded devices are more robust for applications than the DRIE-type structures and are a significant improvement over SiO_2 and Si_xN_y for comparison. Finally, if we consider the loss in a-Si:H reported in Ref. [1], tan δ varies from 4×10^{-4} at lower electric field (low internal power) to 3×10^{-6} at higher electric field (higher internal power). Our control experiment devices have better loss at low power and better loss at high power if we can assume TLS states are saturated by readout power and tan $\delta = 1/Q_i$ holds . In contrast, the wafer bonded devices have about double the loss at low power as a-Si:H and a factor of 3-17 higher at higher power if TLS are saturated. We aim to confirm that we can produce such high quality a-Si:H in a future control experiment for comparison and aim to continue to improve our wafer bonding process.

4. Conclusions

We have measured the loss tangent, tan δ, and internal quality factor, Q_i, for two test devices at RF frequencies between 2-4 GHz. One device used a wafer bonding technique to robustly produce superconducting Nb LC resonators on both sides of a c-Si device layer 5 μm thick, while the other device, our control test device, built similar structures using a less robust method with a DRIE process to test if wafer bonding degraded the performance. We find that the performance based on tan δ and Q_i outperform traditional amorphous dielectrics like SiO_2 and Si_xN_y in both cases, but the performance is reduced in the wafer bonded device compared to the control test device. The wafer bonded device outperforms a-Si:H, but the wafer bonded device does not. Future work will try to identify the extra loss in the wafer bonded device to improve the performance to meet the promise of low loss that c-Si offers for superconducting detectors.

Acknowledgements

The research was carried out at the Jet Propulsion Laboratory, California Institute of Technology, under a contract with the National Aeronautics and Space Administration. Copyright 2016 California Institute of Technology. U.S. Government sponsorship acknowledged.

References

[1] Mazin B A, Sank D, McHugh S, Lucero E A, Merrill A, Gao J, Pappas D, Moore D and Zmuidzinas J 2010 *Applied Physics Letters*, **96** 102504
[2] Weber S J, Murch K W, Slichter D H, Vijay R and Siddiqi I 2011 *Applied Physics Letters*, **98** 172510
[3] Kuo C L, Bock J J, Bonetti J A, Brevik J, Chattopadthyay G, Day P K, Golwala S, Kenyon M, Lange A E, LeDuc H G, Nguyen H, Ogburn R W, Orlando A, Transgrud A, Turner A, Wang G and Zmuidzinas J 2008 *Proc. of SPIE* **7020** 70201I
[4] Sayers J, *et al* 2014 *Proc. of SPIE* **9153** 915304
[5] Shirokoff E, *et al* 2012 *Proc. of SPIE* **8452** 84520R
[6] Gao J *The Physics of Superconducting Microwave Resonators* 2008 California Institute of

Technology Thesis
[7] Patel A, Brown A-D, Hsieh W-T, Stevenson T, Moseley S H, U-yen K, Ehsan N, Barrentine E, Manos G, and Wollack E J 2013 *IEEE Trans. On App. Superconductivity* **23** 2400404
[8] Sold under the name Cyclotene by the Dow Chemical Co.
[9] Denis K L, *et al* 2009 *AIP Conference Proceedings* **1185** 371
[10] Siegel S R, *A Multiwavelength Study of the Intracluster Medium and the Characterization of the Multiwavelength Sub/millimeter Inductance Camera* 2015 California Institute of Technology Thesis

IOP Conf. Series: Journal of Physics: Conf. Series **834** (2017) 012007 doi:10.1088/1742-6596/834/1/012007

Silicon Germanium Cryogenic Low Noise Amplifiers

J.C. Bardin, S. Montazeri, and Su-Wei Chang

Department of Electrical and Computer Engineering, University of Massachusetts Amherst, Marcus 201, 100 Natural Resources Road, Amherst, Massachusetts, 01003-9292, USA

E-mail: jbardin@engin.umass.edu

Abstract. Silicon germanium heterojunction bipolar transistors have emerged in the last decade as an excellent option for use in cryogenic low noise amplifiers. This paper begins with a review of the critical developments that have led to today's cryogenic low noise amplifiers. Next, recent work focused on minimizing the power consumption of SiGe cryogenic amplifiers is presented. Finally, open issues related to the cryogenic noise properties of SiGe HBTs are discussed.

1. Introduction

Microwave cryogenic low noise amplifiers (LNAs) are an enabling element in a number of scientific systems and are widely used in fields ranging from radio astronomy to quantum computing. The first such amplifiers were developed in the early eighties at the National Radio Astronomy Observatory (NRAO). While the initial impetus for research in this area was the desire to realize solid-state amplifiers that could serve as follow-on gain blocks for parametric amplifiers, it was quickly realized these devices—which demonstrated noise temperatures below 10 K and 30 K at 1.5 GHz and 10.7 GHz, respectively—could be used to replace parametric amplifiers altogether, resulting in a significant reduction in system complexity [1, 2, 3, 4].

High carrier mobility is important for a low-noise field-effect transistor and a limitation to the MESFET technology used in these early proof-of-concept-amplifiers is that the channel of a MESFET is doped, which reduces mobility. This problem is resolved in high electron mobility transistors (HEMTs), which were just becoming available in the early eighties. Initial experiments carried out between 1983 and 1984 showed the performance of HEMTs improved dramatically with cryogenic cooling [5, 6]. Encouraged by these results, a collaborative effort to develop HEMTs appropriate for high-performance cryogenic low noise amplifiers was initiated between NRAO and General Electric. As a result of this highly successful program, devices capable of achieving noise temperatures of 0.65 K/GHz were demonstrated in quarter micron InP technology [7, 8, 9]. As HEMT processes improved in subsequent years, so did the cryogenic noise performance of these devices. Today, wideband amplifiers based upon HEMTs are commercially available from Low Noise Factory in Sweden (e.g., [10, 11, 12]) and Cosmic Microwave Technology in California, USA (e.g., [13, 14]).

Over the past decade, silicon germanium (SiGe) heterojunction bipolar transistors have emerged as a viable alternative for the implementation of cryogenic low noise amplifiers. These devices have been developed as part of commercial silicon technology platforms, which have benefits such as large industry investment in process development, synthesizable CMOS logic, and high yield. Thus, SiGe BiCMOS technologies are quite attractive for use in future scientific

IOP Conf. Series: Journal of Physics: Conf. Series **834** (2017) 012007 doi:10.1088/1742-6596/834/1/012007

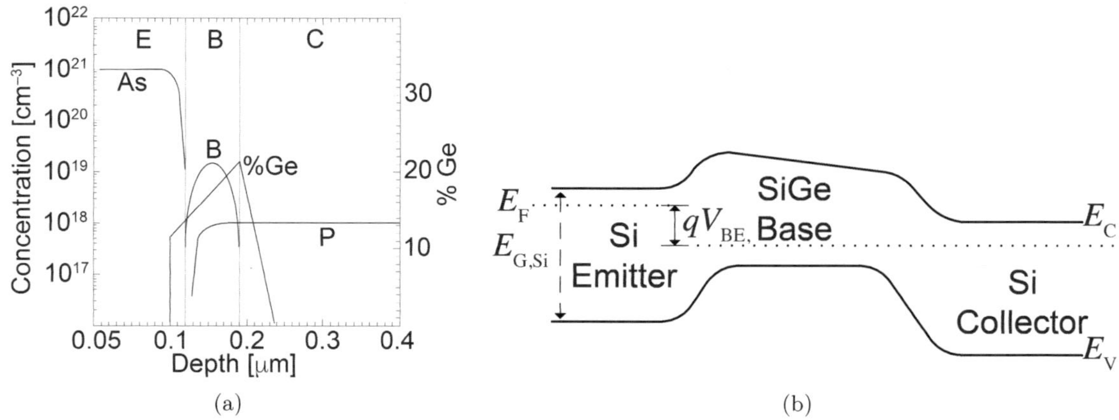

Figure 1. (a) Typical doping and Ge profile of a SiGe HBT. (b) Conceptual band diagram of a SiGe HBT.

sensor systems. In this paper, key developments in cryogenic SiGe LNAs will be reviewed and recent developments in ultra-low-power cryogenic SiGe LNAs will be presented. The paper will conclude with a discussion of open issues related to the cryogenic performance of SiGe HBTs.

2. A Brief History of SiGe Cryogenic LNAs

The idea of a heterojunction bipolar transistor was first proposed by Kroemer in 1957 [15], but it was not until the late eighties that CMOS-compatible SiGe HBTs were demonstrated [16]. Interest in using these devices in cryogenic applications grew shortly thereafter, but before diving into the historical details, it is worth briefly reviewing the properties which make SiGe HBTs attractive for use in these applications. The doping profile and basic band structure of a SiGe HBT appear in Fig. 1. The key improvement in a SiGe HBT over a silicon BJT is the SiGe base alloy which results in a reduced bandgap in the base region compared to that of the collector and emitter regions. Compared to a silicon homojunction bipolar transistor with an otherwise identical composition, the collector current and DC current gain of a SiGe HBT are enhanced by a factor of $\gamma_{\mathrm{DC}} = \exp\{\Delta E_{\mathrm{G0}}/kT_{\mathrm{a}}\}$, where k is Boltzmann's constant, T_{a} is the ambient temperature, $\Delta E_{\mathrm{G0}} = E_{\mathrm{G,Si}} - E_{\mathrm{G,SiGe}}(0)$, $E_{\mathrm{G,Si}} \approx 1.12\,\mathrm{eV}$ is the bandgap of silicon, and $E_{\mathrm{G,SiGe}}(0)$ is the bandgap of the SiGe alloy on the base side of the base–emitter space charge region. In the context of device design, this enhancement of the DC current gain is important since it allows for the design of a device which simultaneously has high DC current gain and low base resistance. In fact, to reduce the base resistance, doping levels well beyond the Mott transition are typically used in the base region, meaning that carrier freeze out in the base at cryogenic temperatures is not an issue [17]. In addition, the Ge content is typically graded across the base region, with increasing Ge towards the collector side of the base. This leads to a sloped conduction band (see Fig. 1(b)), which speeds the carriers as they drift towards the collector. The net result is that the unity current gain cutoff frequency (f_t) of the device is enhanced by a factor of $\Delta E_{\mathrm{G,grade}}/kT_{\mathrm{a}}$, where $\Delta E_{\mathrm{G,grade}} = \Delta E_{\mathrm{G,SiGe}}(0) - \Delta E_{\mathrm{G,SiGe}}(t)$ and $\Delta E_{\mathrm{G,SiGe}}(t)$ is the bandgap at the base side of the base–collector space charge region. Before returning to our historical discussion, it is worth emphasizing that, from the discussion above, both β and f_{t} would be expected to improve tremendously with cryogenic cooling.

Much of the initial research in the cryogenic performance of SiGe HBTs was carried out by Cressler and collaborators at IBM. This work was initially motivated by a desire to improve the switching speeds of emitter coupled logic (ECL). In 1991, this team reported the first measurements of SiGe circuits at liquid nitrogen temperatures [18]. In this initial measurement, the authors observed a marginal increase in f_t, but found that the switching speed of 28 ps obtained using a test device was similar to that observed at room temperature. However, shortly thereafter they found that, by designing a device for operation at cryogenic temperatures, it was feasible to achieve improved values of switching speed, β, and f_t when the device was cooled to 77 K [19]. Significant research in the design of SiGe devices for cryogenic temperatures continued over the next years with both numerical TCAD studies [20] as well as experimental studies in which sophisticated devices were designed with operation at liquid nitrogen temperatures in mind [21, 22, 23].

Today's cryogenic low noise amplifiers must operate at temperatures well below 77 K, where carrier freeze-out and other higher order effects can become significant. The first measurements of SiGe transistors at liquid helium temperatures (4.2 K) were carried out by the IBM team in 1995 [24]. They found that the device did in fact work, but, in comparison to what would be expected from drift-diffusion equations, they found that the transconductance at a given current was lower than expected. Shortly thereafter, this reduced collector current ideality factor was explained as non-equilibrium transport [25, 26].

The first commercial SiGe technology was reported in 1996 [27], and since then, the vast majority of cryogenic SiGe research has been focused on using commercial processes, which are optimized for room temperature operation. The basic cryogenic operation of these technologies has been studied extensively and it now known that state-of-the-art devices operate well at temperatures below 1 K [17] and, with cryogenic cooling, they experience typical increases in β and f_t of approximately 10 x and 50%, respectively [28]. In fact, the best cryogenic results reported for β and f_t are an impressive 80,000 and 798 GHz, respectively [28, 29]. To understand why these improvements are beneficial from a noise perspective, it will be helpful to briefly review the factors that determine the noise performance of a bipolar transistor.

The minimum noise temperature of a bipolar transistor can be expressed as a function of frequency as

$$T_{\mathrm{MIN}} \approx n_c T_a \sqrt{\frac{1}{\beta}\left(1 + 2\frac{g_{\mathrm{m}} R_{\mathrm{B}}}{n_{\mathrm{c}}}\right) + 2\frac{g_{\mathrm{m}} R_{\mathrm{B}}}{n_{\mathrm{c}}}\left(\frac{f}{f_{\mathrm{t}}}\right)^2}, \tag{1}$$

where T_{a} is the ambient temperature, $\beta = I_{\mathrm{C}}/I_{\mathrm{B}}$ is the DC current gain, g_{m} is the intrinsic transconductance, R_{B} is the base resistance, f_{t} is the unity current gain cutoff frequency, $n_{\mathrm{c}} = qI_{\mathrm{C}}/kT_{\mathrm{a}}g_{\mathrm{m}}$ is the collector current ideality factor, q is the charge of an electron, and k is Boltzmann's constant. Thus, if no device properties were to change with cryogenic cooling, the achievable noise performance would improve proportionally as the physical temperature is reduced. However, as described above, significant improvements in both β and f_{t} are typically observed with the cryogenic cooling of a SiGe HBT. Moreover, the base resistance tends to be weakly dependent upon temperature [28], so, as long as the ideality factor is only weakly temperature dependent and a bias consistent with a fixed transconductance is employed, one may deduce that a dramatic improvement in noise performance may be expected when a SiGe HBT is cooled to cryogenic temperatures.

Despite the potential impact of SiGe devices on the field of cryogenic low noise amplifiers, the initial growth of research in this area was somewhat slow. In 2003, cryogenic noise measurements were carried out on a device with a room temperature f_{t} of 200 GHz. Through on-wafer measurements in an open-cycle cryostat, it was found that the minimum noise temperature was roughly 21 K$\pm$8 K at 85 K physical temperature [30]. Similar measurements were later carried out on other device technologies [31, 32, 33, 34], but due to the major challenge associated with the measurement of on-wafer noise temperature measurements with accuracy on the order of $\pm$1 K, no direct measurement of the noise performance in the low GHz frequency range has been carried out.

In 2007, the first serious attempt to use SiGe HBTs in an ultra-low-noise cryogenic amplifier was reported [35]. In this work, a simple method to predict the low-frequency cryogenic noise parameters of a SiGe HBT based upon DC measurements was presented. An IBM (now Global Foundries) 8HP device was then characterized and a narrowband LNA with a noise temperature of approximately 2.5 K was demonstrated at a physical temperature of 15 K. This result was consistent with theory and within 1 K of the best reported noise temperature of an amplifier in this frequency range.

Following this initial success, the noise performance of the IBM 8HP technology was further investigated and a detailed modeling effort was reported at the IEEE International Microwave Symposium the following year [36]. The key result of this paper was that it appeared feasible to realize high-gain cryogenic SiGe HBT amplifiers with noise temperatures as low as 1 K and 6 K at 1 and 10 GHz, respectively. To further explore the capabilities of commercial SiGe technologies, a comprehensive study was carried out in which seven different SiGe HBT technologies were compared in terms of noise performance at 18, 50, 77, 200, and 300 K [28]. As part of this study, the detailed properties of transistors fabricated by IBM, ST, IHP, and TowerJazz were studied and the theoretical limitations for each technology were determined as a function of frequency. Despite the fact that none of these technology platforms were designed with cryogenic operation in mind, the performance was consistently found to improve with cooling. However, it was also discovered that the low-injection collector current ideality factor consistently increased with cooling below 77 K, and for each technology the low-injection product of $n_c T_a$ appeared to settle to a constant value at low enough physical temperatures. The potential mitigation of this effect is an open research topic that will be discussed further in 4.1.

At the same time as these basic device studies, work towards using SiGe HBT technology for practical systems was also underway. In 2008, it became clear that the IF bandwidth of hot electron bolometer (HEB) mixer based THz receivers could be greatly extended if a wideband IF amplifier could be interfaced to an HEB mixer without the use of an isolator. To this end, a proof-of-concept discrete transistor amplifier employing resistive feedback to achieve input matching was developed. The amplifier was subsequently integrated with an HEB receiver [37, 38]. This work was later extended to produce an amplifier that relied on packaged off-the-shelf SiGe transistors, and thus could be replicated as needed for use in various scientific systems [39]. Other research focused on demonstrating amplifiers with the lowest possible noise. In 2009, an impedance matched amplifier with a noise temperature below 2 K was reported [28]. This amplifier employed experimental transistors fabricated by ST and its noise performance is believed to be the lowest reported to date for an impedance matched amplifier covering the 1–3 GHz frequency range.

An advantage of SiGe technology platforms is the capacity for integration and a key research area that has also been explored is the realization of MMIC cryogenic SiGe LNAs. The first high-performance cryogenic SiGe MMIC LNA was reported in 2009 [40]. This amplifier employed resistive feedback to achieve an input match from 0.1–5 GHz, with the lower frequency limit determined by an off-chip coupling capacitor. The chip measured just 0.6 mm × 0.5 mm (see Fig. 2(a)) and the packaged noise performance was found to be better than 5 K over the majority of the frequency range (see Figs. 2(b) and 2(c)). This amplifier was the first high-performance microwave cryogenic LNA that achieved noise and impedance match over a microwave frequency range that extended well below 1 GHz.

Following the demonstration of proof-of-concept 50 Ω amplifiers, follow on research was carried out with the goal of implementing broadband 270 Ω differential amplifiers for integration with self-complimentary feeds. A number of designs were implemented, an example of which appears in Fig. 3. This particular design was reported in [41] and, from DC–4 GHz, it provided a gain of around 30 dB, input and output return losses better than 10 dB, and a common-mode rejection ratio better than 30 dB. Using the novel measurement technique described in [41], the

IOP Conf. Series: Journal of Physics: Conf. Series **834** (2017) 012007 doi:10.1088/1742-6596/834/1/012007

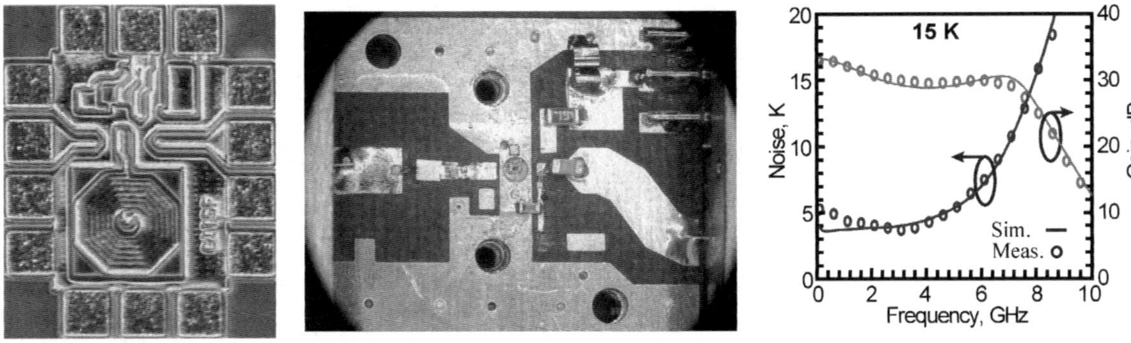

(a) (b) (c)

Figure 2. 0.1–5 GHz cryogenic low noise amplifier. (a) Die photo. The chip measures approximately 0.6 mm × 0.5 mm.(b) Photograph of packaged chip. The die is mounted within a via hole. (c) Noise and gain of the amplifier at a physical temperature of 15 K. The noise was measured to be 5 K or lower from 0.1–5 GHz. Reproduced with permission from [40]. ©2009 IEEE.

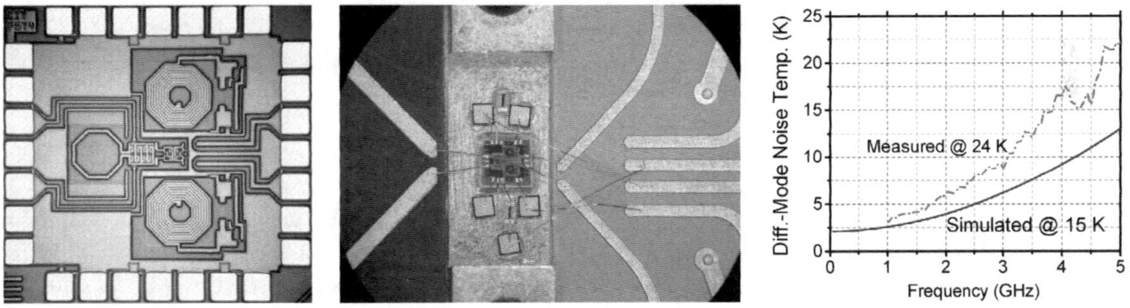

(a) (b) (c)

Figure 3. DC–4 GHz cryogenic 270 Ω differential amplifier. (a) Die photograph, (b) package photograph depicting the differential interconnects, and (c) measured and modeled noise. The noise measurement method was reported in [41] and was somewhat error prone due to a complicated de-embedding procedure and limited gain in the configuration employed for characterization. Reproduced with permission from [41]. ©2010 IEEE.

differential noise temperature was measured at a physical temperature of 24 K and the results appear in Fig. 3(c). Additional fully differential amplifier designs were later reported in [28].

3. Ultra-Low-Power SiGe LNAs

Recent research in SiGe cryogenic MMIC LNA design has been largely driven by a desire to improve the scalability of THz SIS focal plane arrays and other systems with severe power constraints. The ultimate limit to the scaling of a cryogenic receiver system lies in the ability to cool the active electronics. For SIS based systems, the IF LNA must be at a physical temperature of 4 K. Since typical cryocoolers can provide a maximum of 1.5 W heat lift at 4 K [42], the total power consumed by cryogenic amplifiers is limited to a fraction of this. To enable the implementation of kilapixel-scale sideband-separating dual-polarization cameras, the power consumption of each IF amplifier must be reduced to a couple hundred microwatts. This represents about a factor of ten improvement over the current state-of-the-art low-power

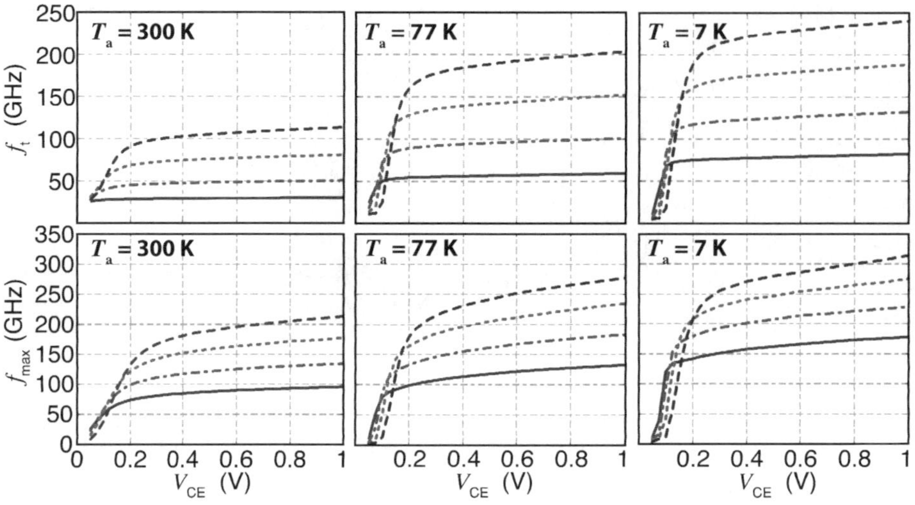

Figure 4. Unity current gain cutoff frequency/maximum frequency of oscillation of a SiGe HBT versus collector–emitter voltage at 300 K (left), 77 K (center), 7 K (right). Data are plotted for current densities of 0.22 mA/μm^2 (solid blue line), 0.46 mA/μm^2 (green dash-dot line), 0.92 mA/μm^2 (red dotted line), and 1.67 mA/μm^2 (purple dashed line). Reproduced with permission from [44]. ©2016 IEEE.

commercial cryogenic LNA, which consumes 4 mW [10].

The initial work in the area of low-power cryogenic SIGe LNAs was reported in 2012 by Russell and Weinreb, who demonstrated a MMIC feedback amplifier that consumed 2 mW while providing a gain of 17 dB and a noise temperature of around 10 K from 2–4 GHz [43]. While DC coupled resistive feedback is a convenient method to supply bias to the base of the transistor while realizing a broadband input match, from the vantage point of power dissipation, it is not necessarily the optimum way to implement a cryogenic low-noise amplifier. The base–emitter turn-on voltage for a typical SiGe HBT operating at cryogenic temperatures is just above 1 V and, if self-biasing is used, the collector must be at a slightly higher potential. Thus, the collector current is the sole degree of freedom available in reducing power consumption if such a topology is employed. However, T_{MIN} is strongly dependent upon current density, so there is a direct trade-off in a self-biased amplifier between noise, power consumption, and bandwidth.

Around 2012, a basic experiment was carried out to determine the minimum collector–emitter voltage that could be used to bias a simple differential pair operating at cryogenic temperatures. It was found that the gain of the circuit was constant for collector–emitter voltages down $\approx$ 150 mV [45]. This unexpected result had a clear implication: by operating SiGe transistors with low collector–emitter voltages and at the nominal bias current for minimum noise, it should be feasible to realize cryogenic LNAs in SiGe technologies without compromising noise performance.

Inspired by this result, a detailed study of the cryogenic performance of SiGe HBTs operating at low collector–emitter voltages was carried out and the results were published in early 2016 [44]. This paper began with the theoretical implications of operating a transistor in weak saturation. Next, the terminal characteristics were presented as a function of collector–emitter voltage and it was found that, contrary to an InP HEMT [46], the small signal and noise performance of a SiGe HBT was insensitive to collector–emitter voltage down to $\approx$200 mV (see Figs. 4 and 5). Interestingly, the linearity was also found to be insensitive to V_{CE} over the same voltage range, provided the collector current density was below 0.5 mA/μm^2 (see Fig. 6). Finally, based upon

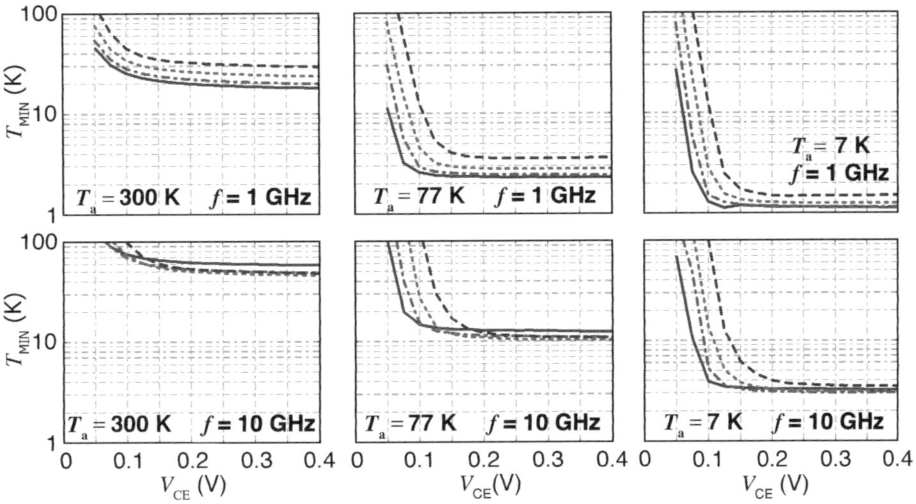

Figure 5. Minimum noise temperature of a SiGe HBT as a function of collector–emitter voltage at 300 K (left), 77 K (center), 7 K (right). The top row is at 1 GHz and the bottom row is at 10 GHz. Data are plotted for current densities of 0.22 mA/μm^2 (solid blue line), 0.46 mA/μm^2 (green dash-dot line), 0.92 mA/μm^2 (red dotted line), and 1.67 mA/μm^2 (purple dashed line). Reproduced with permission from [44]. ©2016 IEEE.

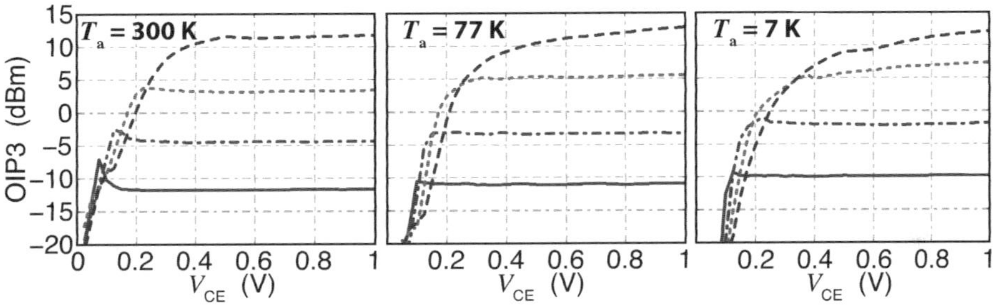

Figure 6. Third order output intermodulation intercept at 3 GHz for a SiGe HBT as a function of collector–emitter voltage at 300 K (left), 77 K (center), and 7 K (right). Data are plotted for current densities of 0.22 mA/μm^2 (solid blue line), 0.46 mA/μm^2 (green dash-dot line), 0.92 mA/μm^2 (red dotted line), and 1.67 mA/μm^2 (purple dashed line). Reproduced with permission from [44]. ©2016 IEEE.

these results, a two stage high-gain cryogenic amplifier that achieved a noise temperature below 5 K from 1.8–3.6 GHz while consuming less than 300 μW was demonstrated.

Following these results, research efforts were shifted to the integration of microwatt SiGe LNAs with SIS mixers, with the goal of demonstrating a scalable THz receiver. In this work, which was a collaborative effort with the Harvard CFA, a discrete transistor amplifier was iteratively designed and characterized with SIS mixers operating in the 200 GHz frequency region. No isolator was employed and the amplifiers were optimized for a noise match to the output impedance of the SIS device. In the first iteration, a 3.3–6 GHz amplifier consuming less than 100 μW and providing an estimated gain of 10 dB was demonstrated [47]. The noise performance of the system using the SiGe amplifier was found to be within 15 K of that of a state-of-the-art setup using the same SIS junction.

(a)

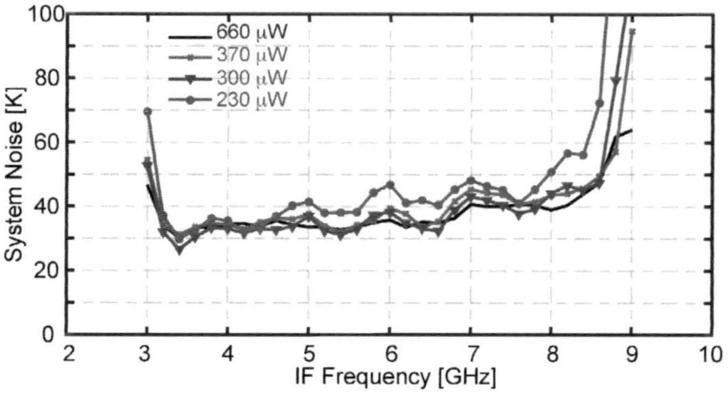

(b)

Figure 7. Ultra-low power THz receiver. (a) Photograph of integrated SiGe LNA and THz SIS receiver. (b) Measured double sideband noise temperature for a local oscillator frequency of 220 GHz. Reproduced with permission from [48]. ©2016 IEEE.

For this initial experiment, a high-power consumption second-stage amplifier was required to provide sufficient gain before exiting the cryostat. Thus, a follow-on effort was carried out to demonstrate that it is feasible to implement a microwatt power consumption SiGe LNA with sufficient gain to directly connect its output to room temperature electronics. An amplifier with approximately 30 dB gain was implemented and the photograph of the amplifier integrated with an SIS module appears in Fig. 7(a) [48]. The measured system noise performance, shown in Fig. 7(b), is both nearly constant for IF amplifier biases as low as 300 μW, and within 5 K of results obtained using a state-of-the-art setup. This is quite encouraging, as it indicates that scalable cryogenic systems employing ultra-low-power SiGe IF amplifiers are practical. To this end, ongoing work is focused on the implementation of MMIC amplifiers with similar performance.

4. Conclusions and Open Questions

In the last decade, it has been proven that commercial SiGe BiCMOS technology platforms can be used to implement state-of-the-art ultra-low-power cryogenic LNAs. However, there are still several open issues related to minimizing the noise of amplifiers using SiGe HBTs. The paper will be concluded with a brief description of two such areas.

4.1. Can the effective carrier temperature be improved?

By introducing the effective electron temperature $T_{\text{eff,elec}} = n_c T_a$, equation (1) can be rewritten as

$$T_{\text{MIN}} \approx \sqrt{\frac{T_{\text{eff,elec}}^2}{\beta} + 2T_a T_0 R_B \frac{I_C}{V_0}\left(\frac{1}{\beta} + \left(\frac{f}{f_t}\right)^2\right)}, \tag{2}$$

where $T_0 = 290$ K is the reference temperature and $V_0 = kT_0/q$. The effective electron temperature is greater than the ambient temperature and is related to the fact that transport at cryogenic temperatures is no longer purely diffusive. Instead, at cryogenic temperatures, the mean free scattering path length for minority carriers in the base is comparable to the base thickness and transport is quasi-ballistic [25, 26]. Thus, the effective electron temperature describes the excess energy of the ballistic carriers. Fig. 8 shows the extracted values of $T_{\text{eff,elec}}$ as a function of physical temperature for several state-of-the-art SiGe technologies. These data were

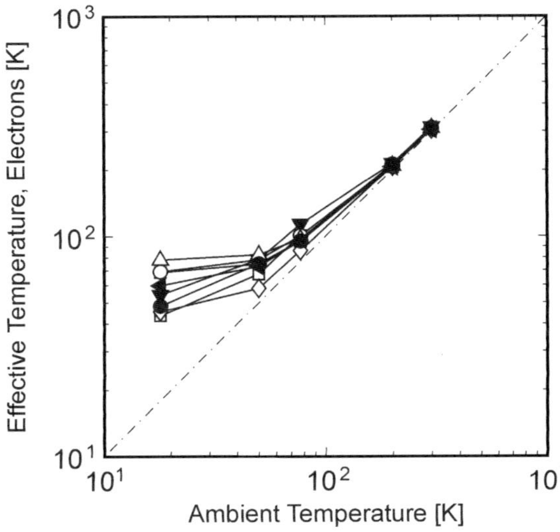

Figure 8. (a) Effective electon and (b) effective hole temperatures as determined from the low-injection slopes of the collector and base currents. State-of-the-art devices from IBM, IHP, ST, TowerJazz, and NXP are included in the figures. Reproduced with permission from [28].

obtained from Gummel curve slopes at current-collector densities several orders of magnitude below the onset of medium injection, and thus are unaffected by self heating. Interestingly, as the temperature tends to zero, the effective electron temperature appears to settle to a constant value, which varies significantly between different device technologies. Moreover, as the current density is increased, the minimum achievable value of $T_{\rm eff}$ is also observed to become larger. Since the zero temperature limit of $T_{\rm MIN}$ is $T_{\rm eff,elec}/\sqrt{\beta}$, it is important to understand the physical limit of $T_{\rm eff,elec}$. Thus, a study of the relationship between device structure and the low-temperature limit of $T_{\rm eff}$ is recommended to try and determine the fundamental limit to this parameter.

4.2. Is shot-noise correlation significant at cryogenic temperatures?

A second open issue has to do with the frequency dependence of the noise parameters of SiGe HBTs at cryogenic temperatures. It is well known that the full shot-noise model of an HBT should include correlation between the base and collector shot-noise currents. This is due to the fact that, rather than being generated in the base–collector junction, the shot noise observed at the collector is actually due to electrons injected across the base–emitter junction. It can be shown that the complete shot-noise model can be written as [49]

$$\overline{|i_{\rm n,b}|^2} = 2qI_{\rm B} + 4qI_{\rm C}\left(1 - \exp\{j\omega\tau_{\rm n}\}\right), \tag{3}$$

$$\overline{|i_{\rm n,c}|^2} = 2qI_{\rm C}, \tag{4}$$

and

$$\overline{i^*_{\rm n,b}i_{\rm n,c}} = 2qI_{\rm C}\left(\exp\{-j\omega\tau_{\rm n}\} - 1\right), \tag{5}$$

where $\tau_{\rm n}$ is a frequency-independent delay term describing the noise transit time to the collector and is a fraction of the overall transit delay ($\tau_{\rm f}$). At room temperature, this delay term is typically equal to around 65% of $\tau_{\rm f}$ [49, 50, 51] and is particularly important for frequencies above $\approx f_{\rm t}/3$ [52].

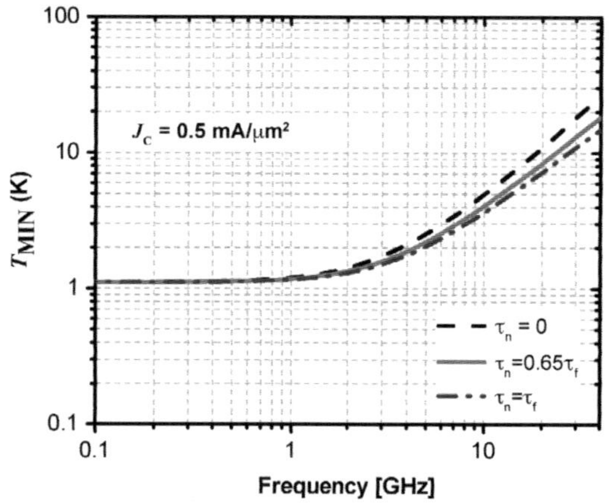

(a)

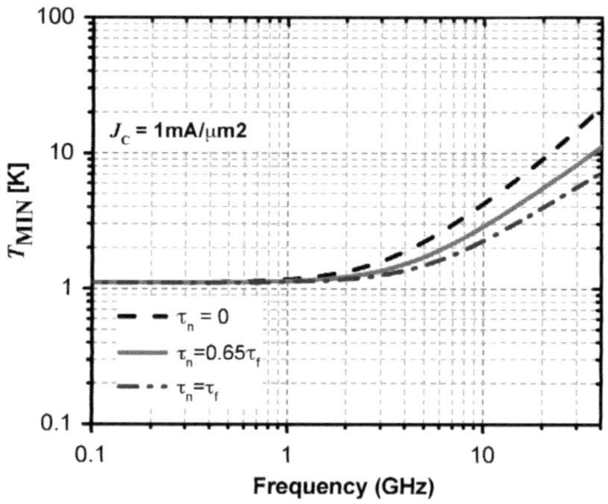

(b)

Figure 9. Simulated minimum noise temperature for an IBM 8HP transistor at 18 K for different noise delay values. Model parameters from the study reported in [44] were employed. The bias points are (a) $0.5\,\mathrm{mA}/\mu\mathrm{m}^2$ and (b) $1\,\mathrm{mA}/\mu\mathrm{m}^2$. The values of f_t for the HBT at these two bias points were found to be 130 and 185 GHz, respectively.

Most cryogenic SiGe HBT noise modeling efforts have neglected shot-noise correlation ($\tau_n = 0$) [28, 36, 43, 44], since determining τ_n requires an error-prone high-frequency cryogenic noise measurement. It is important to note that assuming $\tau_n = 0$ is actually a pessimistic assumption, as the correlated part of the base shot-noise current can actually be leveraged to cancel part of the collector shot-noise current, thereby reducing the overall noise. To emphasize this point, simulated values of T_{MIN} appear in Fig. 9 for current densities of 0.5 and $1\,\mathrm{mA}/\mu\mathrm{m}^2$ and for τ_n equal to 0, $0.65\tau_f$, and τ_f. Interestingly, the effect of shot-noise correlation appears to be quite significant when incorporated into the cryogenic models, especially at higher current densities. However, these optimistic results should be taken with a note of caution; transport in a SiGe HBT at deep cryogenic temperatures is quasi-ballistic rather than diffusive. Thus, there is no reason to believe that the relationship between τ_f and τ_n determined at room temperature could be directly applied at cryogenic temperatures. Thus, further work is required to better understand the effect, if any, that shot-noise correlation has at cryogenic temperatures.

Acknowledgements
This work was partially supported by NSF CAREER grant ECCS-1351744.

References
[1] Weinreb S 1980 *IEEE Transactions on Microwave Theory and Techniques* **28** 1041–1054
[2] Weinreb S, Fenstermacher D L and Harris R W 1982 *IEEE Transactions on Microwave Theory and Techniques* **30** 849–853
[3] Williams D, Lum W and Weinreb S 1980 *Microwave Journal* **23** 73–76
[4] Tomassetti G, Weinreb S and Wellington K 1981 *Electronics Letters* **17** 949–950
[5] Joshin K, Mimura T, Ninori M, Yamashita Y, Kosemura K and Saito J 1983 Noise perfomance of microwave hemt *Microwave Symposium Digest, 1983 IEEE MTT-S International* pp 563–565 ISSN 0149-645X
[6] Laviron M, Delagebeaudeuf D, Rochette J F, Tung P N, Delescluse P, Chevrier J and Linh N T 1984 *IEEE Transactions on Electron Devices* **31** 1967–1967 ISSN 0018-9383
[7] Chao P C, Smith P M, Mishra U K, Palmateer S C, Hwang J C M, Pospieszalski M, Brooks T and Weinreb S 1985 *IEEE Transactions on Electron Devices* **32** 2528–2528 ISSN 0018-9383
[8] Pospieszalski M W, Weinreb S, Chao P C, Mishra U K, Palmateer S C, Smith P M and Hwang J C M 1986 *IEEE Transactions on Electron Devices* **33** 218–223 ISSN 0018-9383

[9] Duh K H G, Pospieszalski M W, Kopp W F, Ho P, Jabra A A, Chao P C, Smith P M, Lester L F, Ballingall J M and Weinreb S 1988 *IEEE Transactions on Electron Devices* **35** 249–256 ISSN 0018-9383

[10] 2010 LNF-LNC4_8A 4-8 GHz cryogenic low noise amplifier Data sheet Low Noise Factory http://www.lownoisefactory.com/products/exband1/lnc-4-8-ghz/

[11] LNF-LNC4_8A data sheet URL http://www.lownoisefactory.com/index.php/download_file/view/19/100/

[12] LNF-LNC7_10A data sheet URL http://www.lownoisefactory.com/index.php/download_file/view/21/85/

[13] 2012 CITCRYO1-12A cryogenic HEMT low noise amplifier Data sheet Cosmic Microwave Technology URL http://cosmicmicrowavetechnology.com/wp-content/uploads/2016/05/CITCRYO1-12DDataSheet12.pdf

[14] CIT416 Datasheet URL http://cosmicmicrowavetechnology.com/wp-content/uploads/2016/05/CIT416DataSheetSep3_2014.pdf

[15] Kroemer H 1957 *Proceedings of the IRE* **45** 1535–1537

[16] King C A, Hoyt J L, Gronet C M, Gibbons J F, Scott M and Turner J 1989 *IEEE Electron Device Letters* **10** 52–54 ISSN 0741-3106

[17] Najafizadeh L, Adams J S, Phillips S D, Moen K A, Cressler J D, Aslam S, Stevenson T R and Meloy R M 2009 *IEEE Electron Device Letters* **30** 508–510

[18] Cressler J D, Comfort J H, Crabbe E F, Patton G L, Lee W, Sun J Y, Stork J M C and Meyerson B S 1991 *IEEE Electron Device Letters* **12** 166–168 ISSN 0741-3106

[19] Cressler J D, Comfort J, Crabbé E, Sun J C and Stork J M 1992 An epitaxial emitter cap, sige-base bipolar technology with 22 ps ecl gate delay at liquid nitrogen temperature *VLSI Technology, 1992. Digest of Technical Papers. 1992 Symposium on* (IEEE) pp 102–103

[20] Richey D, Cressler J and Jaeger R 1994 *Le Journal de Physique IV* **4** C6–127

[21] Cressler J D, Comfort J H, Crabbe E F, Patton G L, Stork J M C, Sun J Y C and Meyerson B S 1993 *IEEE Transactions on Electron Devices* **40** 525–541 ISSN 0018-9383

[22] Cressler J D, Crabbe E F, Comfort J H, Stork J M C and Sun J Y C 1993 *IEEE Transactions on Electron Devices* **40** 542–556 ISSN 0018-9383

[23] Cressler J D, Crabbe E F, Comfort J H, Sun J Y C and Stork J M C 1994 *IEEE Electron Device Letters* **15** 472–474 ISSN 0741-3106

[24] Joseph A J, Cressler J D and Richey D M 1995 *IEEE Electron Device Letters* **16** 268–270 ISSN 0741-3106

[25] Richey D M, Joseph A J, Cressler J D and Jaeger R C 1995 Evidence for nonequilibrium base transport in Si and SiGe bipolar transistors at cryogenic temperatures *Bipolar/BiCMOS Circuits and Technology Meeting, 1995., Proceedings of the 1995* pp 35–38

[26] Richey D M, Joseph A J, Cressler J D and Jaeger R C 1996 *Solid-State Electronics* **39** 785–789

[27] Harame D L and Meyerson B S 2001 *IEEE Transactions on Electron Devices* **48** 2555–2567

[28] Bardin J C 2009 *Silicon-germanium heterojunction bipolar transistors for extremely low-noise applications* Ph.D. thesis California Institute of Technology

[29] Chakraborty P S, Cardoso A S, Wier B R, Omprakash A P, Cressler J D, Kaynak M and Tillack B 2014 *IEEE Electron Device Letters* **35** 151–153

[30] Banerjee B, Venkataraman S, Lu Y, Nuttinck S, Heo D, Chen Y J E, Cressler J D, Laskar J, Freeman G and Ahlgren D 2003 Cryogenic performance of a 200 ghz sige hbt technology *Bipolar/BiCMOS Circuits and Technology Meeting, 2003. Proceedings of the* pp 171–173 ISSN 1088-9299

[31] Banerjee B, Venkataraman S, Lu Y, Liang Q, Lee C H, Nuttinck S, Heo D, Chen Y J, Cressler J D, Laskar J *et al.* 2005 *IEEE Transactions on Electron Devices* **52** 585–593

[32] Pruvost S, Delcourt S, Telliez I, Laurens M, Bourzgui N E, Danneville F, Monroy A and Dambrine G 2005 *IEEE electron device letters* **26** 105–108

[33] Banerjee B, Venkataraman S, Lee C H and Laskar J 2007 Broadband noise modeling of sige hbt under cryogenic temperatures *2007 IEEE Radio Frequency Integrated Circuits (RFIC) Symposium* pp 765–768 ISSN 1529-2517

[34] Waldhoff N, Danneville F, Dambrine G, Geynet B and Chevalier P 2009 Investigation of sige hbt potentialities under cryogenic temperature *2009 Proceedings of the European Solid State Device Research Conference* (IEEE) pp 121–124

[35] Weinreb S, Bardin J C and Mani H 2007 *IEEE Transactions on Microwave Theory and Techniques* **55** 2306–2312

[36] Bardin J C and Weinreb S 2008 Experimental cryogenic modeling and noise of sige hbts *Microwave Symposium Digest, 2008 IEEE MTT-S International* (IEEE) pp 459–462

[37] Pütz P, Honingh C, Justen M, Jacobs K, Bardin J, Mani H and Weinreb S 2008 Integration of if amplifiers with nbtin sheb mixers *Ninteenth International Symposium on Space Terahertz Technology* vol 1 p 416

[38] Pütz P, Schultz M, Honingh C, Justen M, Jacobs K, Bardin J, Jones G, Mani H and Weinreb S 2009 System performance of NbTiN THz SHEB waveguide mixers and cryogenic sige lna *Twentieth International Symposium on Space Terahertz Technology* vol 1 p 161

[39] Weinreb S, Bardin J, Mani H and Jones G 2009 *Review of Scientific Instruments* **80** 044702

[40] Bardin J C and Weinreb S 2009 *IEEE Microwave and Wireless Components Letters* **19** 407–409

[41] Bardin J C and Weinreb S 2010 A dc-4 ghz 270Ω differential SiGe low-noise amplifier for cryogenic applications *Microwave Integrated Circuits Conference (EuMIC), 2010 European* (IEEE) pp 186–189

[42] Rdk 415d capacity map Data sheet Sumitomo Cryogenics http://www.shicryogenics.com//wp-content/uploads/2012/11/RDK-415D_Capacity_Map.pdf

[43] Russell D and Weinreb S 2012 *IEEE Transactions on Microwave Theory and Techniques* **60** 1641–1648

[44] Montazeri S, Wong W T, Coskun A H and Bardin J C 2016 *IEEE Transactions on Microwave Theory and Techniques* **64** 178–187

[45] Gupta D, Bardin J C, Inamdar A, Dayalu A, Sarwana S, Ravindran P, Chang S W, Coskun A H and Sadrabadi M G 2013 *IEEE Transactions on Applied Superconductivity* **23** 1701806–1701806

[46] Schleeh J, Alestig G, Halonen J, Malmros A, Nilsson B, Nilsson P A, Starski J P, Wadefalk N, Zirath H and Grahn J 2012 *IEEE Electron Device Letters* **33** 664–666 ISSN 0741-3106

[47] Montazeri S, Grimes P K, Tong C Y E and Bardin J C 2016 *IEEE Transactions on Terahertz Science and Technology* **6** 133–140

[48] Montazeri S, Grimes P K, Tong C Y E and Bardin J C *IEEE Transactions on Applied Superconductivity* Accepted

[49] Niu G, Cressler J D, Zhang S, Ansley W E, Webster C S and Harame D L 2001 *IEEE Transactions on Electron Devices* **48** 2568–2574

[50] Rudolph M and Heymann P 2007 On compact hbt rf noise modeling *2007 IEEE/MTT-S International Microwave Symposium* (IEEE) pp 1783–1786

[51] Ramirez-Garcia E, Aniel F, Enciso-Aguilar M A and Zerounian N 2012 Intrinsic transit times and noise transport time study of Si/SiGe: C heterejunction bipolar transistors *Microwave Integrated Circuits Conference (EuMIC), 2012 7th European* (IEEE) pp 175–178

[52] Yau K H K, Chevalier P, Chantre A and Voinigescu S P 2011 *IEEE Transactions on Microwave Theory and Techniques* **59** 1983–2000 ISSN 0018-9480

Institute of Physics
Dirac House, Temple Back
Bristol BS1 6BE UK

ISSN: 1742-6588
ISBN 978-1-5108-4042-3

9 781510 840423